Nonlinear Functi

Nonlinear Functional Analysis

Nonlinear Functional Analysis

Rajendra Akerkar

Narosa Publishing House
New Delhi   Madras   Bombay   Calcutta
London

Rajendra Akerkar
Department of Computer Studies
Chh. Shahu Central Institute of Business Education & Research
University Road, Kolhapur, India

Copyright © 1999 Narosa Publishing House

NAROSA PUBLISHING HOUSE
6 Community Centre, Panchsheel Park, New Delhi 110 017
22 Daryaganj, Prakash Deep, Delhi Medical Association Road, New Delhi 110 002
35-36 Greams Road, Thousand Lights, Madras 600 006
306 Shiv Centre, D.B.C. Sector 17, K.U. Bazar P.O., New Mumbai 400 705
2F-2G Shivam Chambers, 53 Syed Amir Ali Avenue, Calcutta 700 019
3 Henrietta Street, Covent Garden, London WC2E 8LU, UK

All rights reserved. No part of this publication may be reproduced, stored in a
retrieval system or transmitted in any form or by any means, electronic,
mechanical, photocopying, recording or otherwise, without the prior permission
of the publishers.

All export rights for this book vest exclusively with Narosa Publishing House.
Unauthorised export is a violation of Copyright Law and is subject to legal action.

ISBN   81-7319-230-8

Published by N.K. Mehra for Narosa Publishing House, 6 Community Centre,
Panchsheel Park, New Delhi 110 017 and printed at Replika Press Pvt. Ltd.,
Delhi 110 040 (India).

To my parents
Ashalata Akerkar
and
Arvind Akerkar

To my parents
Ashobao Merhur
and
Arpna Asceme

PREFACE

This book presents the central ideas of applicable functional analysis in a vivid and straightforward fashion with a minimum of fuss and formality.

The book was developed while teaching an upper-division course in non-linear functional analysis. My intention was to give the background for the solution of nonlinear equations in Banach Spaces, and this is at least one intention of applicable functional analysis. This course is designed for a one-semester introduction at post-graduate level. However, the material can easily be expanded to fill a two semester-course.

To clarify what I taught, I wrote down each delivered lecture. The prerequisites for this text are basic theory on Analysis and Linear Functional Analysis. Any student with a certain amount of mathematical maturity will be able to read the book.

The material covered is more or less prerequisite for the students doing research in applicable mathematics. This text could thus be used for an M.Phil. course in the mathematics.

The preparation of this manuscript was possible due to the excellent facilities available at the Technomathematics Research Foundation, Kolhapur. I thank my colleagues and friends for their comments and help.

I specially thank Mrs. Achala Sabne for the excellent job of preparing the camera ready text.

Most of all, I would like to express my deepest gratitude to Rupali, my wife, in whose space and time this book was written.

R. AKERKAR

CONTENTS

Preface   *v*

1. Contraction   **1**

  1.1 Banach's Fixed Point Theorem   1
  1.2 The Resolvent Operator   9
  1.3 The Theorem of the Local Homeomorphism   11

2. Differential Calculus in Banach Spaces   **17**

  2.1 The Derivative   17
  2.2 Higher Derivatives   28
  2.3 Partial Derivatives   36

3. Newton's Method   **39**

4. The Implicit Function Theorem   **47**

5. Fixed Point Theorems   **55**

  5.1 The Brouwer Fixed Point Theorem   55
  5.2 The Schauder Fixed Point Theorem   60

6. Set Contractions and Darbo's Fixed Point Theorem   **65**

  6.1 Measures of Noncompactness   65
  6.2 Condensing Maps   72

7. The Topological Degree   **77**

  7.1 Axiomatic Definitions of the Brouwer Degree in R^n 77
  7.2 Applications of the Brouwer Degree   80
  7.3 The Leray-Schauder Degree   87
  7.4 Borsuk's Antipodal Theorem   92
  7.5 Compact Linear Operators   99

x Contents

8. Bifurcation Theory   **105**

    8.1 An Example   105
    8.2 Local Bifurcation   110
    8.3 Bifurcation and Stability   116
    8.4 Global Bifurcation   123

9. Exercises and Hints   **129**

    References   **153**
    Index   **155**

Chapter 1

CONTRACTION

1.1 Banach's Fixed Point Theorem

Let $(X, d), (Y, d)$ be metric spaces. A mapping $F : X \to Y$ is said to be **Lipschitz continuous**, if there exists a constant $k > 0$, such that for all $x_1, x_2 \in X$

$$d(F(x_1), F(x_2)) \leq k.d(x_1, x_2).$$

F is called a **contraction**, if for all $x_1, x_2 \in X, x_1 \neq x_2$

$$d(F(x_1), F(x_2)) < d(x_1, x_2).$$

F is called a **strict or a k-contraction**, if F is Lipschitz continuous with a Lipschitz constant $k < 1$.

If $X \subset Y, F : X \to Y$, then $\hat{x} \in X$ is called a **fixed point** of F, if $F(\hat{x}) = \hat{x}$.

If an equation

$$H(x) = y \qquad (1.1)$$

is to be solved, where $H : U \to X$ is a continuous mapping from a subset U of a normed space X into X, then this equation can be transformed in a fixed point problem :

1

Let $T : X \to X$ be an injective (linear) operator, then (1.1) is equivalent to

$$TH(x) = Ty$$
$$x = x - TH(x) + Ty$$

hence

$$x = F(x) \qquad (1.2)$$

where $F(x) = x - TH(x) + Ty$.

The (unique) fixed point x of (1.2) is a (the unique) solution of (1.1), since T is injective. T can be chosen, such that some fixed point principles are applicable. Now we will start with the most important fixed point theorem.

Theorem 1.1 *(Banach's Fixed Point Principle)*
Let X be a complete metric space. Let $F : X \to X$ be a k-contraction with $0 < k < 1$, i.e.

$$\forall\ x_1, x_2 \in X\ \ d(F(x_1), F(x_2)) \le k.d(x_1, x_2).$$

Then the following hold
    1^0 *There exists a fixed point $\hat{x}$ of F.*

    2^0 *This fixed point is unique.*

    3^0 *If $x_0 \in X$ is arbitrarily chosen, then the sequence (x_n), defined by $x_n = F(x_{n-1})$ converges to $\hat{x}$.*

    4^0 *For all n the error estimate is true*

$$d(x_n, \hat{x}) \le \frac{k}{1-k} d(x_n, x_{n-1}) \le \frac{k^n}{1-k} d(x_1, x_0).$$

Contraction

Proof :

For $x_0 \in X$ we have

$$d(x_{n+1}, x_n) = d(F(x_n), F(x_{n-1})) \leq k.d(x_n, x_{n-1})$$
$$\leq ... \leq k^n.d(x_1, x_0)$$

and

$$d(x_{n+j+1}, x_n) \leq d(x_{n+j+1}, x_{n+j}) + ... + d(x_{n+1}, x_n)$$
$$\leq (k^{j+1} + ... + k).d(x_n, x_{n-1})$$
$$\leq \frac{k}{1-k}.d(x_n, x_{n-1})$$
$$\leq \frac{k^n}{1-k}.d(x_1, x_0).$$

Since $k < 1$, the sequence (x_n) is a Cauchy sequence. Since X is complete, $\lim x_n = \hat{x}$ exists. By continuity of F, we have

$$F(\hat{x}) = \lim F(x_n) = \lim x_{n+1} = \hat{x},$$

hence $\hat{x}$ is a fixed point of F.

If $\tilde{x}$ is a fixed point of F, then

$$d(\tilde{x}, \hat{x}) = d(F(\tilde{x}), F(\hat{x})) \leq k.d(\tilde{x}, \hat{x})$$

implies $\tilde{x} = \hat{x}$ (k is less than 1!) and we obtain the error estimates by

$$\lim_{j \to \infty} d(x_{n+j-1}, x_n) = d(\hat{x}, x_n) \leq \frac{k}{1-k}.d(x_n, x_{n-1})$$
$$\leq \frac{k^n}{1-k}.d(x_1, x_0).$$

$\square$

This theorem meets all requirements for a useful mathematical statement: Existence, Uniqueness, Construction and Error Estimate.

4   Nonlinear Functional Analysis

If not necessarily F itself, but almost all iterates

$$F^n = FoF^{n-1}$$

are k_n-contractions, we obtain the following result.

Theorem 1.2 *Let X be a complete metric space, for $F : X \to X$ we assume:*

There exists a sequence (k_n) of positive reals, such that for all $x, y \in X$

$$d(F^n x, F^n y) \leq k_n.d(x, y)$$
$$\Sigma k_n < \infty.$$

Then F has a unique fixed point $\hat{x}$, and $\hat{x} = \lim x_n = \lim F^n(x_0)$ with

$$d(x_n, \hat{x}) \leq \sum_{j \geq n} k_j.d(x_1, x_0).$$

Proof :

This proof is analogous to the proof of Theorem 1.1.

$$
\begin{aligned}
d(x_{n+j+1}, x_n) &\leq d(x_{n+j+1}, x_{n+j}) + ... + d(x_{n+1}, x_n) \\
&\leq d(F^{n+j} x_1, F^{n+j} x_0) + ... + d(F^n x_1, F^n x_0) \\
&\leq (k_{n+j} + ... + k_n).d(x_1, x_0).
\end{aligned}
$$

Thus, (x_n) is a Cauchy sequence. Let $\hat{x} = \lim x_n$, then $F(\hat{x}) = \lim F(x_n) = \lim x_{n+1} = \hat{x}$, i.e. $\hat{x}$ is fixed point; if $\tilde{x}$ is a fixed point of F, so $\tilde{x}$ is a fixed point for all F^n, hence

$$d(F^n \hat{x}, F^n \tilde{x}) \leq k_n.d(\hat{x}, \tilde{x}),$$

implies $\tilde{x} = \hat{x}$, since $k_n < 1$ for almost all n and

$$d(\hat{x}, x_n) \leq \lim_{j \to \infty} d(x_{n+j+1}, x_n) \leq \sum_{j \geq n} k_j.d(x_1, x_0). \qquad \square$$

Contraction

If F is just a contraction, then F does not necessarily have a fixed point:

Let $X = [0, \infty)$ and $F : X \to X$ be defined by

$$F(x) = x + \frac{1}{x+1}.$$

$F(x) = x + \frac{1}{x+1} \neq x$, but

$$F(x) - F(y) = F'(\xi)(x - y) = \left[1 - \frac{1}{(1+\xi)^2}\right](x - y)$$

i.e. $1 - \frac{1}{(1+\xi)^2} < 1$, thus, if $x \neq y$,

$$|F(x) - F(y)| < |x - y|.$$

If we additionally assume that (X, d) is a compact metric space, then we obtain the following result.

Theorem 1.3 *Let X be a compact metric space, $F : X \to X$ a contraction. Then F has a unique fixed point and $\hat{x}$ with $\hat{x} = \lim x_n, \quad x_n = F(x_{n-1}), x_0 \in X$.*

Proof :

Since X is compact, the sequence $(F(x_n))$ has a convergent subsequence $(F(x_{n_j}))$. Let

$$\hat{x} = \lim_{j \to \infty} F(x_{n_j}),$$

then

$$F(\hat{x}) = \lim_{j \to \infty} F(x_{n_j+1}).$$

If $\hat{x} \neq F(\hat{x})$, there exist disjoint closed neighbourhoods U of $\hat{x}$ and V of $F(\hat{x})$. The mapping

$$\rho : U \times V \to \mathcal{R}, \quad \rho(x, y) = \frac{d(F(x), F(y))}{d(x, y)}$$

Nonlinear Functional Analysis

is continuous, and attains its maximum $k < 1$. Let $p \in \mathcal{N}$, such that for $j \geq p$

$$F(x_{n_j}) \in U, \quad F(x_{n_j+1}) \in V.$$

Then

$$d(F(x_{n_j+2}), F(x_{n_j+1})) \leq k.d(F(x_{n_j+1}), F(x_{n_j}))$$

and

$$d(F(x_n), F(x_{n+1})) \leq d(F(x_m), F(x_{m+1}))$$

for $n > m$.

Hence for $j > p$

$$\begin{aligned}
d(F(x_{n_j}), F(x_{n_j+1})) &\leq d(F(x_{n_{j-1}+1}), F(x_{n_{j-1}+2})) \\
&\leq k.d(F(x_{n_{j-1}}), F(x_{n_{j-1}+1})) \leq \cdots \\
&\leq k^{j-p+1}.d(F(x_{n_p+1}), F(x_{n_p+2})) \\
&\leq k^{j-p}.d(F(x_{n_p}), F(x_{n_p+1})).
\end{aligned}$$

Therefore

$$d(\hat{x}, F(\hat{x})) = \lim_{j \to \infty} d(F(x_{n_j}), F(x_{n_j+1})) = 0.$$

This contradiction shows that $\hat{x}$ is a fixed point of F. The uniqueness follows from the contraction property. Finally, we will show $\hat{x} = \lim x_n$. This follows from

$$d(\hat{x}, F(x_{n_j+1})) = d(F^1(\hat{x}), F^1(F(x_{n_j}))) \leq d(\hat{x}, F(x_{n_j})).$$

$\square$

The following example shows that there exist contractions on compact spaces, which are not strict:

Let $F : [-1, 0] \to [-1, 0]$ be defined by $F(x) = x + x^2$.

Then

$$F(x) - F(y) = x + x^2 - y - y^2 = x - y + (x + y)(x - y)$$
$$= (1 + x + y)(x - y).$$

If $x \neq y$, then $|1 + x + y| < 1$, but there is no $k < 1$, such that $|F(x) - F(y)| \leq k|x - y|$.

As an application of Banach's fixed point theorem we will consider the following nonlinear **Volterra integral equation**

$$x(t) - \int_0^t k(t, \tau, x(\tau))d\tau = y(t). \tag{*}$$

Assume the function

$$k : [0, 1] \times [0, 1] \times \mathcal{R} \to \mathcal{R}$$

is continuous and fulfills the following Lipschitz condition: there is a $\gamma > 0$, such that for all $t, \tau \in [0, 1]$, $r, s \in \mathcal{R}$

$$|k(t, \tau, r) - k(t, \tau, s)| \leq \gamma|r - s|.$$

Then the mapping F, defined by

$$F(x)(t) = \int_0^t k(t, \tau, x(\tau))d\tau$$

maps $C[0, 1]$ into $C[0, 1]$. As the complete metric space we choose $(X, d) = (C[0, 1], d_\gamma)$ with

$$d_\gamma(x_1, x_2) = \max_{0 \leq t \leq 1} |x_1(t) - x_2(t)|e^{-2\gamma}.$$

Then $F : X \to X$ and F is a $\frac{1}{2}$-contraction.

$$d_\gamma(F(x_1), F(x_2)) = \max |F(x_1(t)) - F(x_2(t))|e^{-2\gamma}$$
$$\leq \max \int_0^t |k(t, \tau, x_1(\tau)) - k(t, \tau, x_2(\tau))|d\tau . e^{-2\gamma t}$$

$$\leq \quad \max \int_0^t \gamma |x_1(\tau) - x_2(\tau)| e^{-2\gamma\tau} d\tau . e^{-2\gamma t}$$

$$\leq \quad d_\gamma(x_1, x_2) . \gamma . \max_{0 \leq t \leq 1} \int_0^t e^{2\gamma\tau} d\tau . e^{-2\gamma t}$$

$$\leq \quad d_\gamma(x_1, x_2) . \gamma . \frac{1}{2\gamma} \max e^{-2\gamma t} (e^{2\gamma t} - 1)$$

$$\leq \quad \frac{1}{2} d_\gamma(x_1, x_2).$$

By Banach's fixed point theorem we have the following result.

The Volterra integral equation $(*)$ has for every continuous function y a unique continuous solution x.

Especially, we obtain the theorem of Picard - Lindelof :

The initial value problem

$$y' = f(t, y), \quad y(0) = \eta \qquad \left(\ * \ \right)$$

with Lipschitz continuous f has a unique solution, since $\left(\ * \ \right)$ is equivalent to

$$y(t) = \eta + \int_0^t f(\tau, y(\tau)) d\tau.$$

In general operator F is defined on a subset U of the complete metric space X or is a k-contraction only on a subset of X. In such cases it is an additional problem to find a subset $U_0 \subset U$ with the properties: F maps U_0 into U_0, and U_0 itself is a complete metric space. A sufficient condition for such a situation is described in the following result.

Theorem 1.4 *Let* (X, d) *be a complete metric space. Let* $U \subset X$ *and* $F : U \to X$ *be a* k - *contraction with* $k < 1$. *Let* $x_1 \in U, x_2 = F(x_1) \in U, r = \frac{k}{1-k} d(x_1, x_2)$. *Let the closed ball* $U_0 := B(x_2, r) \subset U$. *Then:*

1. *F maps U_0 into U_0.*

Contraction

9

2. *F has a unique fixed point $\hat{x} \in U_0$ and the sequence*
   $(x_n) = (F(x_{n-1}))$ *converges to $\hat{x}$.*

Proof :

The closed subset $U_0 = B(x_2, r)$ is complete. Let $x \in U_0$. Then

$$
\begin{aligned}
d(F(x, x_2)) &= d(F(x), F(x_1)) \le k.d(x, x_1) \\
&\le k(d(x, x_2) + d(x_2, x_1)) \\
&\le k(\frac{k}{1-k} + 1).d(x_1, x_2)) \\
&\le r
\end{aligned}
$$

hence $F : U_0 \to U_0$ and Theorem 1.1 applies.

$\square$

1.2 The Resolvent Operator

Let X be a Banach space and $U \subset X$. Let $F : U \to X$ be a continuous mapping. Let $V \subset X$ be a subset, such that for all $y \in V$ the equation

$$
x - F(x) = y
$$

has a unique solution x. Then x can be represented by

$$
x = y - R(y)
$$

and the mapping $R : V \to X$ is said to be the **resolvent operator** to F.

In the case, where F is a contraction with Lipschitz constant $k < 1$, the resolvent operator exists and is Lipschitz continuous, too.

Theorem 1.5 *Let X be a Banach space and $F : X \rightarrow X$ be Lipschitz continuous with Lipschitz constant $k < 1$. Then the resolvent operator R to F exists and has Lipschitz constant $\frac{k}{1-k}$.*

Proof :

For every $y \in X$ the equation

$$x - F(x) = y \qquad (1.1)$$

has a unique solution, since the operator $F_0 : X \rightarrow X$ defined by

$$F_0(x) = F(x) + y$$

is a k-contraction and has a fixed point $x \in X$. So a mapping $G : X \rightarrow X$ with $G(y) = x$ is defined. Let $R(y) = y - G(y)$, then

$$x = y - R(y) \qquad (1.2)$$

is the unique solution of equation (1.1). Let $y_1, y_2 \in X$ and $x_j - F(x_j) = y_j$. Then

$$\begin{aligned}
\|R(y_1) - R(y_2)\| &= \|x_1 - y_1 - (x_2 - y_2)\| = \|F(x_1) - F(x_2)\| \\
&\leq k\|x_1 - x_2\| \\
&\leq k\|y_1 - R(y_2) - (y_2 - R(y_1))\| \\
&\leq k\|y_1 - y_2\| + k\|R(y_1) - R(y_2)\|
\end{aligned}$$

hence

$$\|R(y_1) - R(y_2)\| \leq \frac{k}{1-k} \cdot \|y_1 - y_2\|$$

$\square$

The representation

$$x = y - R(y)$$

of the solution x of equation (1.1) shows that properties of R determine the structure of the solution. We will illustrate this fact by the following result.

Contraction

Theorem 1.6 *Let X be a Banach space, $F : U \to X$ a continuous operator and $R : V \to X$ its resolvent opeator. Let $Y \subset X$ be a linear subspace, such that* Range $F \subset Y$. *Then* Range $R \subset Y$.

Proof :

From equations (1.1) and (1.2) it follows that

$$R(y) = -F(x) = -F(y - R(y)).$$

Thus Range R $\subset$ Range $(-F) \subset Y$.

$\square$

Theorem 1.7 *Let $\Omega \subset \mathcal{R}^n$ be an open subset, $X = C(\Omega)$. Let $k : \Omega \times \Omega \times \mathcal{R} \to \mathcal{R}$, such that $D_t^\alpha k(., \tau, \xi)$ exists for all $\alpha \in \mathcal{N}^n$ with $|\alpha| \leq \nu, \tau \in \Omega, \xi \in \mathcal{R}$. Then*

$$F : C(\Omega) \to C(\Omega)$$

defined by

$$(Fx)(t) = \int_\Omega k(t, \tau, x(\tau)) d\tau$$

maps $C(\Omega)$ into the subspaces $C^\nu(\Omega) \subset C(\Omega)$ of ν - times differentiable functions. If further the resolvent operator R to F exists, then the solution x of an equation

$$x - F(x) = y$$

$y \in C(\Omega)$, has the property $x - y \in C^\nu(\Omega)$.

1.3 The Theorem of the Local Homeomorphism

The strongest version of the solution of the equation

$$F(x) = y$$

is the construction of the inverse map F^{-1}, such that

$$x = F^{-1}(y).$$

The existence (and construction) of this inverse map in some cases is possible. In this section we will see, that F is (continuous) invertible, if F can be approached by a linear continuous invertible operator T in the following sense.

Theorem 1.8 *Let X, Y be Banach spaces and $U = B(x_0, r)$ a closed ball in X with centre x_0 and radius r. Let $F : U \to Y$ be a mapping, and let $T : X \to Y$ a continuous linear homeomorphism, such that there exist q with $0 < q < 1$, such that for all $u, v \in U$,*

$$||F(u) - F(v) - T(u - v)|| \le q||T(u - v)||. \qquad (*)$$

Then

1^0 *F is injective and continuous in U.*

2^0 *Let $\rho = r(1 - q).||T^{-1}||^{-1}$. The closed ball $V = B(F(x_0), \rho)$ with centre $F(x_0)$ and radius ρ is contained in $F(U)$, i.e. $V \subset F(U)$.*

3^0 *F^{-1} is continuous in V.*

4^0 *Let $U_0 = F^{-1}(V)$. Then U_0 is a closed neighbourhood of x_0, and $F|_{U_0} : U_0 \to V$ is a homeomorphism from U_0 onto V.*

5^0 *For all $y_1, y_2 \in V$ we have the inequality*

$$||F^{-1}y_1 - F^{-1}y_2 - T^{-1}(y_1 - y_2)|| \le p||T^{-1}(y_1 - y_2)||$$

with $p = \frac{q}{1-q}.||T||^2.||T^{-1}||^2$.

Contraction

Proof :

1^0 Let $F(u) = F(v)$ for some $u, v \in U$. Then $(*)$ implies

$$\|T(u - v)\| \leq q\|T(u - v)\|$$

and since $q < 1$, $Tu = Tv$ and $u = v$ by the injectivity of T.

2^0 Let $y \in V$. Then $F(x) = y$ is equivalent to the equations $F(T^{-1}z) = y$ and $Tx = z$, and further to

$$z = G(z)$$

where

$$G(z) = z - F(T^{-1}z) + y.$$

We will see that G is a q-contraction on the closed ball $V_0 = B(Tx_0, r\|T^{-1}\|^{-1})$. Let $z_1, z_2 \in V_0$, $z_j = Tx_j$, then

$$
\begin{aligned}
\|G(z_1) - G(z_2)\| &= \|z_1 - z_2 - F(T^{-1}z_1) + F(T^{-1}z_2)\| \\
&\quad \|Tx_1 - Tx_2 - F(x_1) + F(x_2)\| \\
&\leq q\|T(x_1 - x_2)\| = q\|z_1 - z_2\|,
\end{aligned}
$$

since $\|z_j = Tx_0\| \leq r\|T^{-1}\|^{-1}$ and

$$
\begin{aligned}
\|x_j - x_0\| &= \|T^{-1}(z_j - Tx_0)\| \\
&\leq \|T^{-1}\|.\|z_j - Tx_0\| \\
&\leq \|T^{-1}\|.r.\|T^{-1}\|^{-1} = r,
\end{aligned}
$$

i.e. $x_j \in U$.

This shows that G is a contraction. Now we will see, G maps V_0 into itself. Let $z \in V_0$, $z_0 = Tx_0$, then

$$\begin{aligned}
||G(z) - Tx_0|| &= ||z - F(T^{-1}z) + y - Tx_0|| \\
&\leq ||z - F(T^{-1}z) - (z_0 - F(T^{-1}z_0))|| + \\
&\quad + ||y - F(T^{-1}z_0)||) \\
&\leq q||z - z_0|| + ||y - F(x_0)|| \\
&\leq qr \cdot ||T^{-1}||^{-1} + (1-q)r.||T^{-1}||^{-1} \\
&= r.||T^{-1}||^{-1}.
\end{aligned}$$

$(y \in V$ implies $||y - F(x_0)|| \leq (1-q)r.||T^{-1}||^{-1})$.

Since V_0 is a closed ball in a Banach space, V_0 is complete, and by the fixed point theorem of Banach G has a unique fixed point $\hat{z} \in V_0$. Thus $\hat{x} = T^{-1}\hat{z}$ is the unique solution of $F(x) = y$.

3^0 Let $y_1, y_2 \in V$, $z_1, z_2 \in V_0$ be the fixed points of

$$G_1(z) = z - F(T^{-1}z) + y_1$$

$$G_2(z) = z - F(T^{-1}z) + y_2$$

By the properties in 2^0 we obtain

$$\begin{aligned}
||z_1 - z_2|| &= ||G_1(z_1) - G_2(z_2)|| \\
&\leq ||G_1(z_1) - G_2(z_1)|| \\
&\quad + ||G_2(z_1) - G_2(z_2)|| \\
&\leq ||y_1 - y_2|| + q||z_1 - z_2||,
\end{aligned}$$

hence
$$||z_1 - z_2|| \leq \frac{1}{1-q}.||y_1 - y_2||.$$

For $x_j = F^{-1}(y_j)$ we obtain

$$||x_1 - x_2|| = ||T^{-1}(z_1 - z_2)|| \leq ||T^{-1}||.||z_1 - z_2|| \leq \frac{||T^{-1}||}{1-q}.||y_1 - y_2||$$

Differential Calculus in Banach Spaces

and the continuity of F^{-1} follows from

$$||F^{-1}(y_1) - F^{-1}(y_2)|| \le \frac{||T^{-1}||}{1-q}.||y_1 - y_2||, \ \text{if} \ y_1, y_2 \in V.$$

4^0 Since V is closed, U_0 is closed by continuity of F.

5^0 Let $y_1, y_2 \in V, w = F^{-1}(y_1) - F^{-1}(y_2) - T^{-1}(y_1 - y_2)$.
Then with $x_j = F^{-1}(y_j)$,

$$Tw = T(x_1 - x_2) - F(x_1) + F(x_2)$$

and

$$
\begin{aligned}
||w|| &\le ||T^{-1}||.||Tw|| \le ||T^{-1}||.||F(x_1) - F(x_2) - T(x_1 - x_2)|| \\
&\le ||T^{-1}||.q.||T(x_1 - x_2)|| \\
&\le ||T^{-1}||.||T||.q.||x_1 - x_2|| \\
&\le ||T^{-1}||^2.||T||.\frac{q}{1-q}.||y_1 - y_2|| \\
&\le ||T^{-1}||^2.||T||^2.\frac{q}{1-q}.||T^{-1}.(y_1 - y_2)||
\end{aligned}
$$

This proves 5^0.

$\square$

Chapter 2

DIFFERENTIAL CALCULUS IN BANACH SPACES

2.1 The Derivative

In this chapter, X will be a Banach space, X^* its dual space, i.e. the space of continuous linear functionals. Let $U \subset X$ be an open subset of X, $x_0 \in U$ and $f : U \to \mathcal{R}$ a real valued function. There are several possible meanings to the statement, that $x^* \in X^*$ is the derivative of f at x_0.

Definition 2.1 1^0 x^* *is the Gâteaux derivative of f at x_0,* *iff*

$$\forall \ u \in X, \lim_{\tau \to 0} \frac{1}{\tau}(f(x_0 + \tau u) - f(x_0) - \tau x^*(u)) = 0$$

2^0 x^* *is the Fréchet derivative of f at x_0 iff*

$$\lim_{\|u\| \to 0} \frac{1}{\|u\|}(f(x_0 + u) - f(x_0) - x^*(u)) = 0.$$

In both cases, we denote by $f'(x_0) = x^* \in X^*$ the derivative. Clearly, $2^0 \Rightarrow 1^0$.

If f is *Fréchet* differentiable at x, then f is continuous at x, since

$$\forall \epsilon > 0 \ \exists \delta > 0 \ \forall y \in X \ \|y\| \leq \delta$$
$$\Rightarrow \|f(x+y) - f(x) - x^*(y)\| \leq \epsilon\|y\|$$

and

$$\|f(x+y) - f(x)\| \leq \epsilon\|y\| + \|x^*\|.\|y\|$$

implies the continuity of f at x.

On the other hand, let

$$f : \mathcal{R}^2 \to \mathcal{R}, f(0) = 0, f(\xi,\eta) = \frac{\xi}{\eta}(\xi^2 + \eta^2).$$

Let $x^* = 0$. Then

$$\frac{1}{\gamma}(f(0+\tau y) - f(0) - x^*(y)) = \frac{1}{\tau}.f(\tau y) = \frac{1}{\tau}.\frac{\tau\xi}{\tau\eta}.\tau^2(\xi^2 + \eta^2)$$

hence

$$\lim_{\tau \to 0} \frac{1}{\tau} f(\tau y) = 0$$

and $f(0) = 0$ is the *Gâteaux* derivative of f, but f is not continuous at 0.

This definition can be extended to map between Banach spaces. Let X, Y be Banach spaces, $U \subset X$ an open subset and $F : U \to Y$ a map.

Definition 2.2 1^0 *A continuous linear map $T : X \to Y$ is the Gâteaux derivative of F at $x_0 \in U$ iff*

$$\forall u \in X \ \lim_{\tau \to 0} \frac{1}{\tau}.\|F(x_0 + \tau u) - F(x_0) - \tau Tu\| = 0.$$

2^0 *A linear map $T : X \to Y$ is the Fréchet derivative of F at $x_0 \in U$ iff*

$$\lim_{\|u\| \to 0} \frac{1}{\|u\|}.\|F(x_0 + u) - F(x_0) - Tu\| = 0.$$

Differential Calculus in Banach Spaces

3^0 T *is the weak Gâteaux derivative of F at x_0 iff*

$$\forall u \in X \ \forall x^* \in X^* \ \lim_{\tau} \frac{1}{\tau} x^*(F(x_0 + \tau u) - F(x_0) - \tau Tu) = 0.$$

4^0 T *is the weak Fréchet derivative of F at x_0 iff*

$$\forall x^* \in X^* \ \lim_{||u|| \to 0} \frac{1}{||u||} . x^*(F(x_0 + u) - F(x_0) - Tu) = 0.$$

We denote by $F'(x_0)$ the derivative of F at x_0.

(Weakly) *Fréchet* differentiable maps are (weakly) continuous, but there are weakly continuous maps, which are not continuous, e.g. $F : [0, 1] \to c_0$

$$F(t) = (f_n(t))$$
$$F(0) = 0$$

$$f_n(t) = \begin{cases} 1 & t = \frac{1}{n} \\ 0 & |t - \frac{1}{n}| \geq \frac{1}{2n} \\ \text{linear} & \text{and continuous elsewhere} \end{cases}$$

(a) $\lim_{n \to \infty} f_n(t) = 0$, since $f_n(t) = 0$, if $n \geq \frac{1}{2t}$
    hence $(f_n(t)) \in c_0$.

(b) $||F(0 + h) - F(0)|| = \max_n |f_n(h)| = 1, h = \frac{1}{m}$, hence
    F is not continuous at 0.

(c) $c_0^* = \ell_1$. Let $x^* = (\xi_n) \in \ell_1$. Choose for $\epsilon > 0$ an
    integer n_0, such that

$$\sum_{n > n_0} |\xi_n| \leq \epsilon,$$

then

$$|x^*(F(h))| = |\sum \xi_n f_n(h)| \leq |\sum_{n \leq n_0} \xi_n f_n(h)| + \sum_{n > n_0} |\xi_n|.$$

The finite sum of continuous functions is continuous hence, if $|h| \leq \frac{1}{2n_0}$, then $|x^*(F(h))| \leq \epsilon$, and F is weakly continuous at 0.

20   Nonlinear Functional Analysis

Remark 2.1 *Let $F : U \to Y$ be continuous and Fréchet differentiable at x_0, then $F(x_0)$ is a continuous linear map.*

Let $\epsilon > 0, \ \delta > 0$ such that if $||u|| \leq \delta \leq 1$

$$||F(x_0 + u) - F(x_0)|| \leq \frac{\epsilon}{2}$$

and

$$||F(x_0 + u) - F(x_0) - F'(x_0)u|| \leq \frac{\epsilon}{2}.||u|| \leq \frac{\epsilon}{2}.$$

Then

$$||F'(x_0)u|| \leq \epsilon \ \text{ if } \ ||u|| \leq \delta.$$

This is the continuity of $F'(x_0)$ at the origin, hence everywhere.

Remark 2.2 *If F is Fréchet differentiable at x_0, then there exist $\gamma > 0, \delta > 0$, such that for all $x \in U$*

$$||x - x_0|| \leq \delta \Rightarrow ||F(x) - F(x_0)|| \leq \gamma.||x - x_0||.$$

Let $\varphi_F(x) = F(x) - F(x_0) - F'(x_0)(x - x_0)$, then

$$||F(x) - F(x_0)|| \leq ||F'(x_0)||.||x - x_0|| + ||\varphi_F(x)||.$$

Since, $\lim \frac{\varphi_F(x)}{||x-x_0||} = 0$, there is a $\delta > 0$, such that $||x - x_0|| \leq \delta \Rightarrow ||\varphi_F(x)|| \leq ||x - x_0||$. Thus

$$||F(x) - F(x_0)|| \leq (1 + ||F'(x_0)||).||x - x_0||.$$

Examples

1. The constant mapping $F(x) = a$ for $x \in U$ has the derivative $F'(x) = 0$.

2. Let $F : X \to Y$ be a continuous linear map. Then

$$F(x + u) - F(x) - F(u) = 0.$$

   Hence $F'(x) = F$ for every $x \in X$.

Differential Calculus in Banach Spaces

3. Let $k : [0,1] \times [0,1] \times \mathcal{R} \to \mathcal{R}$ be continuous and $\frac{\partial}{\partial t_3} k(t_1, t_2, t_3)$ be continuous for $t_1, t_2 \in [0,1], t_3 \in \mathcal{R}$. Let $F : C[0,1] \to C[0,1]$ be defined by

$$F(x)(t) = \int_0^1 k(t, \tau, x(\tau)) d\tau,$$

then F is in $x \in C[0,1]$ Fréchet differentiable and $F'(x)$ is the linear operator

$$F'(x) : C[0,1] \to C[0,1]$$

$$F'(x)u(t) = \int_0^1 \frac{\partial}{\partial t_3} k(t, \tau, x(\tau)).u(\tau) d\tau.$$

This can be seen by the following.

Let $t_3 \in \mathcal{R}$, for $\epsilon > 0$ choose $\delta > 0$, such that, if $|h| \leq \delta$, then

$$|k(t, \tau, t_3 + h) - k(t, \tau, t_3) - \frac{\partial}{\partial t_3} k(t, \tau, t_3).h| \leq \epsilon|h|.$$

If $t_3 = x(\tau)$, then

$$|k(t, \tau, x(t) + h) - k(t, \tau, x(\tau)) - \frac{\partial}{\partial t_3} k(t, \tau, x(\tau)) \ . \ h| \leq \epsilon|h|.$$

Let $u \in C[0,1]$ with $|u| \leq \delta$, then $|u(\tau)| \leq \delta$ and

$$\int_0^1 |k(t, \tau, x(\tau) + u(\tau)) - k(t, \tau, x(\tau))$$

$$- \frac{\partial}{\partial t_3} k(t, \tau, x(\tau)), u(\tau)| d\tau$$

$$\leq \epsilon \|u\|.$$

4. Let $f : \mathcal{R} \to \mathcal{R}$ be differentiable and

$$F : C[0,1] \to C[0,1]$$

$$F(x)(t) = f(x(t))$$

Then
$$F'(x) : C[0,1] \to C[0,1]$$
$$F'(x)u(t) = f'(x(t)).u(t)$$

i.e. the *Fréchet* derivative of F is the multiplication operator, which multiplies each continuous function u by $f'ox$.

5. Let X, Y, Z be Banach spaces and the Banach space $X \times Y$ endowed with the norm $||(u,v)|| = \max(||u||, ||v||)$. Let $B : X \times Y \to Z$ be a continuous bilinear map, i.e.

$$||B|| = \sup_{\substack{||x|| \le 1 \\ ||y|| \le 1}} ||B(x,y)|| < \infty.$$

Let $(x_0, y_0) \in X \times Y$. Then

$$||B(u,v)|| \le ||B||.||u||.||v|| \le ||B||.||(u,v)||^2$$

thus

$$\lim_{||(u,v)|| \to 0} \frac{||B(u,v)||}{||(u,v)||} = 0.$$

Therefore B is *Fréchet* differentiable and $B'(x_0, y_0)$ is the linear map

$$B'(x_0, y_0) : X \times Y \to Z$$
$$(u, v) \to B(x_0, v) + B(u, y_0).$$

6. Let X, Y be two isomorphic Banach spaces and $\mathcal{GL}(X, Y)$ the space of all continuous linear operators T from X onto Y, which are continuously invertible. Then the mapping

$$F : \mathcal{GL}(X, Y) \to \mathcal{GL}(Y, X)$$

$$F(t) = T^{-1}$$

is differentiable in every point T_0 and

$$F'(T_0)H = -T_0^{-1} H T_0^{-1}.$$

Differential Calculus in Banach Spaces

Proof :

If $\|H\| < \|T_0^{-1}\|^{-1}$, then

$$\|F(T_0 + H) - F(T_0) - F'(T_0)H\|$$
$$= \|(T_0 + H)^{-1} - T_0^{-1} + T_0^{-1}HT_0^{-1}\|$$
$$= \|(I + T_0^{-1}H)^{-1}.T_0^{-1} - T_0^{-1} + T_0^{-1}HT_0^{-1}\|$$
$$= \|\sum_{j=0}^{\infty}(-T_0^{-1}H)^j.T_0^{-1} - T_0^{-1} + T_0^{-1}HT_0^{-1}\|$$
$$= \|\sum_{j=2}^{\infty}(-T_0^{-1}H)^j\|.\|T_0^{-1}\| \leq \|H\|^2.\frac{\|T_0^{-1}\|^3}{1 - \|T_0^{-1}H\|}$$

hence

$$\lim_{H \to 0} \frac{\|F(T_0 + H) - F(T_0) - T_0^{-1}HT_0^{-1}\|}{\|H\|} = 0.$$

$\square$

Theorem 2.1 *(Chain Rule)*

Let X, Y, Z be Banach spaces, $U \subset X, V \subset Y$ open subsets. Let $F : U \to Y, G : V \to Z$ be continuous mappings, such that $F(U) \subset V$, $y_0 = F(x_0)$ and let the Fréchet derivatives $F'(x_0)$ and $G'(y_0)$ exist. Then GoF is Fréchet differentiable and

$$(GoF)'(x_0) = G'(y_0)oF'(x_0).$$

(The derivative of the composition is the composition of the derivative).

Proof :

Let $H = GoF, y = F(x)$ and

$$\varphi_F(x) = F(x) - F(x_0) - F'(x_0)(x - x_0)$$

$$\varphi_G = G(y) - G(y_0) - G'(y_0)(y - y_0)$$

24   Nonlinear Functional Analysis

then

$$\begin{aligned}
\varphi_G &= H(x) - H(x_0) - G'(y_0)(F(x) - F(x_0)) \\
&= H(x) - H(x_0) - G'(y_0)[F'(x_0)(x - x_0) + \varphi_F(x)] \\
&= H(x) - H(x_0) - G'(y_0)F'(x_0)(x - x_0) - G'(y_0)\varphi_F(x)
\end{aligned}$$

hence

$$\begin{aligned}
\varphi_H &= H(x) - H(x_0) - G'(y_0)F'(x_0)(x - x_0) \\
&= G'(y_0)\varphi_F(x) + \varphi_G(F(x)).
\end{aligned}$$

Now it suffices to prove

$$\lim_{x \to x_0} \frac{\varphi_H(x)}{||x - x_0||}.$$

Since $G'(x_0)$ is a bounded linear operator

$$||G'(y_0)\varphi_F(x)|| \leq ||G'(y_0)||.||\varphi_F(x)||.$$

F is differentiable, hence

$$\forall \epsilon > 0 \; \exists \delta_1 \; \forall x \in U \; ||x - x_0|| \leq \delta_1 \; \Rightarrow \; ||\varphi_F(x)|| \leq \epsilon.||x - x_0||.$$

G is differentiable, hence

$$\forall \epsilon > 0 \; \exists \delta_2 \; \forall y \in V \; ||y - y_0|| \leq \delta_2$$
$$\Rightarrow ||\varphi_G(y)|| \leq \epsilon.||y - y_0||.$$

Since F is differentiable, we have

$$\exists \gamma > 0 \; \exists \delta_3 > 0 \; \forall x \in U \; ||x - x_0|| \leq \delta_3$$
$$\Rightarrow ||F(x) - F(x_0)|| \leq \gamma.||x - x_0||.$$

Since F is continuous

$$\forall \delta_2 > 0 \; \exists \delta_4 > 0 \; \forall x \in U \; ||x - x_0|| \leq \delta_4$$
$$\Rightarrow ||F(x) - F(x_0)|| \leq \delta_2.$$

Differential Calculus in Banach Spaces

Thus we obtain for $x \in U, ||x - x_0|| \le \min(\delta_1, \delta_3, \delta_4)$

$$
\begin{aligned}
||\varphi_H(x)|| &\le ||G'(y_0)||.||\varphi_F(x)|| + ||\varphi_G(F(x))|| \\
&\le ||G'(y_0)||.\epsilon.||x - x_0|| + \epsilon.||F(x) - F(x_0)|| \\
&\le ||G'(y_0)||.\epsilon.||x - x_0|| + \epsilon.\gamma.||x - x_0|| \\
&\le \epsilon.(||G'(y_0)|| + \gamma).||x - x_0||,
\end{aligned}
$$

hence

$$
\lim \frac{||\varphi_H(x)||}{||x - x_0||} = 0
$$

and

$$
H'(x_0) = G'(y_0)oF'(x_0).
$$

$\square$

Theorem 2.2 *Let $U \subset X$ be an open subset and $F : U \to Y$ be a homeomorphism from U onto the open subset $V \subset Y, V = F(U)$. Let F be Frechet differentiable at $x_0 \in U$, such that $F'(x_0) : X \to Y$ is a linear homeomorphism. Then the inverse mapping $G = F^{-1} : V \to X$ is Fréchet differentiable at $y_0 = F(x_0)$ and*

$$
G'(y_0) = F'(x_0)^{-1}.
$$

(The derivative of the inverse is the inverse of the derivative).

Proof :

If F is differentiable at x_0,

$$
\forall \epsilon > 0 \; \exists \delta > 0 \; \forall x \in U \;\; ||x - x_0|| \le \delta \Rightarrow
$$

$$
||\varphi_F(x)|| = ||F(x) - F(x_0) - F'(x_0)(x - x_0)|| \le \epsilon.||x - x_0||
$$

$$
\le \epsilon.||F'(x_0)^{-1}||.||F'(x_0)(x - x_0)||.
$$

Let $\epsilon^* = \epsilon.||F'(x_0)^{-1}|| < 1$. Then by the theorem of the local homeormorphism (Theorem 1.8(5^0))

$$||G(y) - G(y_0) - F'(x_0)^{-1}(y - y_0)|| \le p(\epsilon^*).||y - y_0||$$

with $p(\epsilon^*) = \frac{\epsilon^*}{1-\epsilon^*}.||F'(x_0)||^2.||F'(x_0)^{-1}||^3$.

This shows, that G is differentiable at y_0 and

$$G'(y_0) = F'(x_0)^{-1}.$$

$\square$

Theorem 2.3 *(Mean Value Theorem)*

Let X be a Banach space, $U \subset X$ an open subset of X. Let $x_0, x_0 + h \in U$, such that the segment

$$[x_0, x_0 + h] = \{x_0 + \tau h, 0 \le \tau \le 1\} \subset U.$$

Let $F : U \to Y$ be continuous and Fréchet differentiable for all $x \in [x_0, x_0 + h]$. Then

$$||F(x_0 + h) - F(x_0)|| \le ||h||. \sup_{x \in [x_0, x_0 + h]} ||F'(x)||.$$

Proof :

Let $y^* \in Y^*$ be a continuous linear functional. The real valued function ψ of the real variable τ

$$\psi(\tau) = y^*(F(x_0 + \tau h)) \quad 0 \le \tau \le 1$$

is differentiable by the chain rule and

$$\begin{aligned}
\psi(1) - \psi(0) &= y^*(F(x_0 + h) - F(x_0)) \\
&= \psi'(\tau_m) = y^*(F'(x_0 + \tau_m h)h), \quad 0 < \tau_m < 1.
\end{aligned}$$

Differential Calculus in Banach Spaces

By the Hahn - Banach theorem we have

$$\|F(x_0 + h) - F(x_0)\| = \sup_{\substack{\|y^*\| = 1 \\ y^* \in Y^*}} y^*(F(x_0 + h) - F(x_0))$$

$$\leq \sup_{0 \leq \tau \leq 1} \|F'(x_0 + \tau h)\| \cdot \|h\|.$$

$\square$

Remarks :

1^0 If dim $X > 1$, then in general there is no equality in the statement of the mean value theorem. Let
$f : \mathcal{R} \to \mathcal{C},$
$f(t) = e^{it}, h = 2\pi$, then
$\quad f(t + h) - f(t) = 0$, but
$\quad f'(t + \tau_m h) = ie^{it + 2\tau_m \pi i} \neq 0.$

2^0 In the situation of Theorem 2.3 we have for
$u \in [x_0, x_0 + h] : \|F(x_0 + h) - F(x_0) - F'(u)h\|$
$\leq \|h\| \sup_{v \in [x_0, x_0 + h]} \|F'(v) - F'(x_0)\|.$
This can be verified by the application of the mean value theorem to the function
$$G(x) = F(x) - F'(u)x.$$
for $u = x_0$.

3^0 The statement 2^0 can be sharpened, if $F'(x)$ is uniformly bounded for all $x \in U$. Let

$$k = \sup_{x \in U} \|F'(x)\| < \infty.$$

Then for all $x, y \in U$

$$\|F(x) - F(y)\| \leq k\|x - y\|.$$

This follows from the equations

$$\varphi(t) = F(x + t(y - x))$$

$$\varphi'(t) = F'(x + t(y - x))(y - x)$$

$$\|F(x) - F(y)\| = \|\varphi(1) - \varphi(0)\| = \|\int_0^1 \varphi'(t)dt\|$$
$$\leq \int \|\varphi'(t)\| dt \leq k\|x - y\|.$$

If additionally F' is Lipschitz continuous in U, i.e.

$$\|F'(y) - F'(x)\| \leq m\|y - x\|$$

then

$$\|F(y) - F(x) - F'(x)(y - x)\| \leq \frac{m}{2}\|y - x\|^2,$$

since

$$\psi(t) = F(x + t(y - x)) - tF'(x)(y - x)$$
$$\psi'(t) = F'(x + t(y - x))(y - x) - F'(x)(y - x)$$

implies that

$$\|F(y) - F(x) - F'(x)(y - x)\| = \|\psi(1) - \psi(0)\|$$
$$= \|\int_0^1 \psi'(t)dt\|$$
$$\leq m\int_0^1 tdt.\|y - x\|^2.$$

2.2 Higher Derivatives

Before going to higher derivatives, we will study multilinear mappings between Banach spaces.

Let $X_1, ..., X_n, Y$ be Banach spaces, and $\mu C(X_1, ..., X_n; Y)$ the vector space of all n-linear mappings

$$M : X_1 \times ... \times X_n \to Y$$

i.e. the space of those mappings which are linear in each variable.

Differential Calculus in Banach Spaces

Proposition 2.1 *A* n - *linear mapping* $M : X_1 \times ... \times X_n \to Y$ *is continuous iff there is a constant* γ, *such that for all* $(x_1, ..., x_n) \in X_1 \times ... \times X_n$

$$||M(x_1, ..., x_n)|| \leq \gamma.||x_1||...||x_n||.$$

Proof :

" $\Leftarrow$". Let $n = 2$; and $(a_1, a_2) \in X_1 \times X_2$. To show that M is continuous in (a_1, a_2), we write

$$M(x_1, x_2) - M(a_1, a_2) = M(x_1 - a_1, x_2) + M(a_1, x_2 - a_2)$$

then

$$||M(x_1, x_2) - M(a_1, a_2)|| \leq \gamma||x_1 - a_1||.||x_2|| + \gamma||a_1||.||x_2 - a_2||$$

$$||(x_1, x_2) - (a_1, a_2)|| = ||x_1 - a_1|| + ||x_2 - a_2|| \to 0$$

implies the continuity of M.

" $\Rightarrow$". Let M be continuous in $(0, 0)$. Then there is a ball

$$B = \{(x_1, x_2) \in X_1 \times X_2, ||x_1|| + ||x_2|| \leq r\}$$

such that for $(x_1, x_2) \in X_1 \times X_2$ we have $||M(x_1, x_2)|| \leq 1$. Let $(x_1, x_2) \neq (0, 0), z_i = \frac{rx_i}{2||x_i||}$, i.e. $(z_1, z_2) \in B, ||M(z_1, z_2)|| \leq 1$. Then

$$M(z_1, z_2) = M\left[\frac{rx_1}{2||x_1||}, \frac{rx_2}{2||x_2||}\right] = \frac{r^2}{4||x_1||.||x_2||}.M(x_1, x_2).$$

Hence

$$\frac{r^2}{4||x_1||.||x_2||}.||M(x_1, x_2)|| \leq 1$$

Nonlinear Functional Analysis

implies

$$\|M(x_1, x_2)\| \leq \frac{4}{r^2}\|x_1\|.\|x_2\|.$$

□

For $M \in \mu C(X_1, ..., X_n; Y)$ we define

$$\|M\| = \sup \{\|M(x_1, ..., x_n)\|, \|(x_1, ..., x_n)\| \leq 1\}.$$

This defines a norm on $\mu C(X_1, ..., X_n; Y)$.

Proposition 2.2 *Let X, Y, Z be Banach space. The space $\mathcal{L}(X, \mathcal{C}(Y, Z))$ and $\mu C(X \times Y; Z)$ are norm isomorphic, if $X \times Y$ is endowed with the norm $\|(x, y)\| = max(\|x\|, \|y\|)$.*

Proof :

Let $M \in \mu C(X \times Y; Z)$. For $x \in X$ we define $M_x \in \mathcal{L}(Y, Z)$ by $M_x(y) = M(x, y)$. The mapping $T_M : X \to \mathcal{L}(Y, Z)$, defined by $T_M x = M_x$ is linear and the linear mapping

$$\phi : M \to T_M$$

is an isomorphism of $\mu C(X \times Y; Z)$ onto $\mathcal{L}(X, \mathcal{L}(Y, Z))$

$$\|\phi(M)\| = \|T_M\| = \sup_{\|x\| \leq 1} \|M_x\| = \sup_{\|x\| \leq 1} \sup_{\|y\| \leq 1} \|M(x, y)\| = \|M\|.$$

If $T \in \mathcal{L}(X, \mathcal{L}(Y, Z))$, then for $x \in X, T_x \in \mathcal{L}(Y, Z)$, i.e. $M : (x, y) \to T_x y$ is bilinear and

$$\|M(x, y)\| = \|T_x y\| \leq \|T_x\|.\|y\|$$

and

$$\|M\| = \sup_{\substack{\|x\| \leq 1 \\ \|y\| \leq 1}} \|T_x y\| = \|T\|.$$

Differential Calculus in Banach Spaces

Together with the above statement this means: ϕ is bijective and $||\phi|| = 1$. $\qquad\square$

Let $U \subset X$ be an open subset of the Banach space X and $F : U \geq Y$. We say that F is continuously differentiable, if $F' : U \to \mathcal{L}(X, Y)$ is continuous. We say that F is twice differentiable at $x_0 \in U$, if F' is continuous and F' is differentiale at x_0. Then we write $F''(x_0)$ and $F'''(x_0) \in \mathcal{L}(X, \mathcal{L}(X, Y))$. Since $\mathcal{L}(X, \mathcal{L}(X, Y))$ can be identified with the space $\mu C(X \times X; Y)$ of bilinear operators, we understand $F'''(x_0)$ as a bilinear operator with

$$(g, h) \to F'(x_0)(g, h).$$

Proposition 2.3 *Let F be twice differentiable at $x_0 \in U$, let $g \in X$. then the map $x \to F'(x)g$ from U into Y is differentiable at x_0, and its derivative is the linear operator*

$$h \to F''(x_0)(h, g).$$

Proof :

The mapping $x \to F'(x)g$ is the composition of $T \to Tg$ (from $(\mathcal{L}(X, Y)$ into Y) and $x \to F'(x)$ (from U into $\mathcal{L}(X, Y)$). By the chain rule we obtain the result. $\qquad\square$

Theorem 2.4 *Let F be twice differentiable at x_0. Then the bilinear mapping $(g, h) \to F'''(x_0)(g, h)$ is symmetric, i.e.*

$$F''(x_0)(g, h) = F''(x_0)(h, g).$$

Proof :

Let $\varphi : [0, 1] \to Y$ be defined by

$$\varphi(\tau) = F(x_0 + \tau g + h) - F(x_0 + \tau g)$$

Nonlinear Functional Analysis

where we assume $||g|| \leq \delta/2$, $||h|| \leq \delta/2$, $B(x_0, \delta) \subset U$. By the mean value theorem

$$||\varphi(1) - \varphi'(0)|| \leq \sup ||\varphi'(\tau) - \varphi'(0)||$$

and by the chain rule

$$\begin{aligned}
\varphi'(\tau) &= (F'(x_0 + \tau g + h) - F'(x_0 + \tau g))g \\
&= (F'(x_0 + \tau g + h) - F'(x_0))g - (F'(x_0 + \tau g) - F'(x_0))g.
\end{aligned}$$

The differentiability of $F'(x)$ implies, that for given $\epsilon > 0$ there exists $\delta^* > 0$, such that, if $||h|| \leq \delta^2/2, ||g|| \leq \delta^*/2$

$$||(F'(x_0 + \tau g + h) - F'(x_0))g - F''(x_0)(\tau g + h, g)|| \leq \epsilon.||\tau g + h||.||g||$$

$$||(F'(x_0 + \tau g) - F'(x_0))g - F''(x_0)(\tau g, g)|| \leq \epsilon.\tau||g||^2.$$

Therefore we obtain

$$||\varphi'(\tau) - F''(x_0)(h, g)|| \leq 2.\epsilon(||g|| + ||h||)^2.$$

Now we obtain

$$\begin{aligned}
&||\varphi(1) - \varphi(0) - F''(x_0)(h, g)|| \\
&\leq ||\varphi(1) - \varphi(0) - \varphi'(0)|| + ||\varphi'(0) - F''(x_0)(h, g)|| \\
&\leq \sup_\tau ||\varphi'(\tau)|| + ||\varphi'(0) - F''(x_0)(h, g)|| \\
&\leq \sup_\tau ||\varphi'(\tau) - F''(x_0)(h, g)|| + ||F''(x_0)(h, g) - \varphi'(0)|| \\
&\quad + ||\varphi'(0) - F''(x_0)(h, g)|| \\
&\leq 3.2.\epsilon(||g|| + ||h||)^2.
\end{aligned}$$

The term

$$\varphi(1) - \varphi(0) = F(x_0 + g + h) - F(x_0 + h) - F(x_0 + g) + F(x_0)$$

Differential Calculus in Banach Spaces 33

is symmetric in g and h, hence

$$||F''(x_0)(h,g) - F''(x_0)(g,h)||$$
$$\leq ||\varphi(1) - \varphi(0) - F''(x_0)(h,g)||$$
$$+||\varphi(1) - \varphi(0) - F''(x_0)(g,h)||$$
$$\leq 12.\epsilon(||g|| + ||h||)^2,$$

whenever $||g|| \leq \delta^*/2, ||h|| \leq \delta^*/2$.

If we choose $||(g,h)|| = \max(||g||, ||h||)$, then we see

$$||F''(x_0)(h,g) - F''(x_0)(g,h)|| \leq 48.\epsilon||(g,h)||^2,$$

this means, that the bilinear map
$(g,h) \to F''(x_0)(h,g) - F''(x_0)(g,h)$ has norm zero; and this is
the symmetry of the second derivative.

$\square$

By induction, we can define k-times differentiable mappings:
Let $F : U \to Y, F$ is said to be k-times differentiable at x_0, if
the derivative of order $k - 1$ is continuous and differentiable at
x_0. We write $F^{(k)}(x_0).F^{(k)}(x_0)$ can be identified with a k - linear
map, $F^{(k)}(x_0) \in \mu C(X \times ... \times X; Y)$.

Example

Let $f : \mathcal{R} \to \mathcal{R}$ be a twice continuously differentiable function. Let $F : C[0,1] \to C[0,1]$ defined by

$$F(x)(t) = f(x(t)).$$

Then

$$F''(x_0)(g, h)(t) = f''(x_0(t)).(g(t).h(t)).$$

We conclude this chapter with the two versions of the theorem
of Taylor.

Theorem 2.5 *Let X, Y be Banach spaces, $U \subset X$ open, x, $x + h \in U$, such that $[x, x + h] \subset U; F : U \to Y$ be p-times differentiable and the derivatives up to the order $p - 1$ continuous. Then*

$$\|F(x + h) - F(x) - F'(x)h - \ldots -$$
$$\frac{1}{(p - 1)!} \cdot F^{(p-1)}(x)(h, \ldots, h)\|$$
$$\leq \frac{\|h\|^p}{p!} \sup_{[x, x+h]} \|F^{(p)}(u)\|.$$

Proof :

Let $y^* \in Y^*$ be a continuous linear functional. The real valued function

$$\psi(\tau) = y^*(F(x + \tau h)), \quad \tau \in [0, 1]$$

is p-times differentiable and

$$\psi^{(k)}(\tau) = y^*(F^{(k)}(x + \tau h)(h, \ldots, h)).$$

The Taylor formula for real functions gives

$$y^*(F(x + h) - \sum_{j=0}^{p-1} \frac{1}{j!} F^{(j)}(x)(h, \ldots, h))$$

$$= \psi(1) - \sum_{j=0}^{p-1} \frac{1}{j!} \psi^{(j)}(0)$$

$$= \frac{1}{p!} \psi^{(p)}(\tau_m), \quad 0 < \tau_m < 1$$

$$= y^*(\frac{1}{p!} F^{(p)}(x + \tau_m h)(h, \ldots, h))$$

Taking sup of y^*, such that $\|y^*\| \leq 1$ and $\tau_m \in [0, 1]$ the proof is complete. $\qquad\square$

Differential Calculus in Banach Spaces

Corollary 2.1 *Let $F : U \to Y$ as in Theorem 2.5, and $x \to F^{(p)}(x)$ be continuous. Then for every $\epsilon > 0$ there is a $\delta > 0$, such that for all $h \in X, ||h|| \leq \delta$*

$$||F(x+h) - \sum_{j=0}^{p} \frac{1}{j!} F^{(j)}(x)(h, ..., h)|| \leq \epsilon.||h||^p.$$

Proof :

Let $p = 2$ and

$$G(x) = F(x) - F(y) - F'(y)x - \frac{1}{2}F''(y)(x, x).$$

Then

$$\begin{aligned} G(x+h) &= F(x+h) - F(y) - F'(y)(x+h) \\ &\quad - \frac{1}{2}F''(y)(x+h, x+h) \\ G'(x)h &= F'(x)h - F'(y)h - F''(y)(x, h) \end{aligned}$$

and

$$G(x+h) - G(x) - G'(x)h = F(x+h) - F(x) - F'(x)h - \frac{1}{2}F'(y)(h, h).$$

By Theorem 2.8 we have

$$||G(x+h) - G(x) - G'(x)h|| \leq \frac{||h^2||}{2} \sup_{u \in [x, x+h]} ||G''(u)||.$$

It is

$$G'''(x)(h, h) = F'''(x)(h, h) - F'''(y)(h, h)$$

thus

$$\sup_u ||G'''(u)|| = \sup_u ||F'''(u) - F'''(y)||$$

and, letting $y = x$

$$\|F(x+h) - F(x) - F'(x)h - \frac{1}{2}F''(x)(h,h)\|$$

$$\leq \frac{\|h\|^2}{2} \sup_{u \in [x,x+h]} \|F''(u) - F''(x)\|.$$

The continuity of F'' guarantees the existence of $\delta > 0$, such that $\|F''(u) - F''(x)\| \leq \epsilon$ if $\|h\| \leq \delta$.

In general, let $G(x) = F(x) - \sum_{j=0}^{p} \frac{1}{j!}F^{(j)}(y)(x,...x)$.

Theorem 2.5, applied to G gives for $y = x$

$$\|F(x+h) - \sum_{j=0}^{p} \frac{1}{j!}F^{(j)}(x)(h,...,h)\|$$

$$\leq \frac{\|h\|^p}{p!} \sup_{u \in [x,x+h]} \|F^{(p)}(u) - F^{(p)}(x)\|.$$

The continuity of $F^{(p)}$ guarantees the existence of $\delta > 0$, such that $\|F^{(p)}(u) - F^{(p)}(x)\| \leq \epsilon$, if $\|u - x\| \leq \delta$ and $\|h\| \leq \delta$. $\qquad \square$

2.3 Partial Derivatives

Let $X = X_1 \times X_2, U_u \subset X_i$ open, $(a_1, a_2) \in U_1 \times U_2 = U$. Let $F : U \to Y$ be differentiable. We define the partial maps

$$x_1 \to F(x_1, a_2), \quad x_2 \to F(a_1, x_2)$$

from U_i into Y and say, that F is partially differentiable in $(a_1, a_2) \in U$, iff the maps

$$x_1 \to F(x_1, a_2) \quad \text{and} \quad x_2 \to F(a_2, x_2)$$

are differentiable in a_i. By $D_j F(a_1, a_2)$ we denote the partial derivative of F with respect to the first (resp. second) coordinate.

Differential Calculus in Banach Spaces

Theorem 2.6 *Let $F : U_1 \times U_2 \to Y$ be a continuous map from the open set $U = U_1 \times U_2 \subset X$ into Y. Then F is continuously differentiable in $(a_1, a_2) \in U$ if and only if F is partially differentiable and the partial derivatives are continuous mappings*

$$(x_1, x_2) \to D_1 F(x_1, x_2)$$
$$U \to \mathcal{L}(X_1, Y)$$

and

$$(x_1, x_2) \to D_2 F(x_1, x_2)$$
$$U \to \mathcal{L}(X_2, Y).$$

The (total) derivative of F in (a_1, a_2) is given by

$$F'(a_1, a_2)(h_1, h_2) = D_1 F(a_1, a_2)h_1 + D_2 F(a_1, a_2)h_2.$$

Proof :

" $\Rightarrow$ " : The mappings $G_1 : x_1 \to F(x_1, a_2)$ is the composition of F and $i_1^{a_2} : x_1 \to (x_1, a_2)$. The derivative of the second map is the linear map $i_1 : h_1 \to (h_1, 0)$. By the chain rule we get

$$\begin{aligned}
D_1 F(a_1, a_2)h_1 = G_1'(a_1)h_1 &= (F o i_1^{a_2})'(a_1)h_1 \\
&= F'(i_1^{a_2}(a_1)) o i_1(h_1) \\
&= F'(a_1, a_2)(h_1, 0)
\end{aligned}$$

and similarly,

$$\begin{aligned}
D_2 F(a_1, a_2)h_2 = G_2'(a_1)h_2 &= (F o i_2^{a_1})'(a_2)h_2 \\
&= F'(i_2^{a_1}(a_2)) o i_2(h_2) \\
&= F'(a_1, a_2)(0, h_2)
\end{aligned}$$

Since $i_1 + i_2 = id$, we obtain

$$\begin{aligned}
F'(a_1, a_2)(h_1, h_2) &= F'(a_1, a_2)(h_1, 0) + F'(a_1, a_2)(0, h_2) \\
&= D_1 F(a_1, a_2)h_1 + D_2 F(a_1, a_2)h_2. \quad (*)
\end{aligned}$$

Nonlinear Functional Analysis

" $\Leftarrow$ " : We will show that for given $\epsilon > 0$ there is a $\delta > 0$, such that, if $||(h_1, h_2)|| \le \delta$, we have

$$\Delta = ||F(a_1 + h_1, a_2 + h_2) - F(a_1, a_2) \\ -D_1 F(a_1, a_2)h_1 - D_2 F(a_1, a_2)h_2|| \le \epsilon.||(h_1, h_2)||.$$

By the differentiability of the partial maps we obtain for $||h_1|| \le \delta_1$

$$||F(a_1 + h_1, a_2) - F(a_1, a_2) - D_1 F(a_1, a_2)h_1|| \le \epsilon.||h_1||,$$

by the mean value theorem we obtain

$$||F(a_1 + h_1, a_2 + h_2) - F(a_1 + h_1, a_2) - D_2 F(a_1 + h_1, a_2)h_2|| \\ \le ||h_2|| \sup_{||z|| \le ||h_2||} ||D_2 F(a_1 + h_1, a_2 + z) - D_2 F(a_1 + h_1, a_2)||$$

by the continuity of $D_2 F$, there is a $\delta_2 > 0$, such that

$$||h_1|| \le \delta_2, ||h_2|| \le \delta_2$$

$$||D_2 F(a_1 + h_1, a_2) - D_2 F(a_1, a_2)|| \le \epsilon$$

and

$$\sup_{||z|| \le \delta_2} ||D_2 F(a_1 + h_1, a_2 + z) - D_2 F(a_1 + h_1, a_2)|| \le \epsilon.$$

Thus we obtain for $||(h_1, h_2)|| \le \min(\delta_1, \delta_2)$

$$\Delta \le 3\epsilon.||(h_1, h_2)||.$$

The continuity of F' follows from $(*)$.

$\square$

Chapter 3

NEWTON'S METHOD

Since it is important to know how fast the convergence of a sequence (x_n) to the limit z really is, one usually introduces the order of convergence as a first asymptotic test.

Definition 3.1 *Let (x_n) be a sequence in the Banach space X with* $\lim x_n = \hat{x}$. *The sequence (x_n) is said to be convergent of order $p > 1$, if there exist positive reals β, γ, such that for all $n \in \mathcal{N}$*

$$\|x_n - \hat{x}\| \le \beta.exp(-\gamma p^n) \ .$$

If the sequence (x_n) has the property, that

$$\|x_n - \hat{x}\| \le \beta q^n, \ \ 0 < q < 1$$

then (x_n) is said to be linearly convergent.

Proposition 3.1 *If $\hat{x}$ is the fixed point of $F : U \to X$ and if there exist reals $p > 1, \beta_p > 0$, such that for all $x \in U$*

$$\|F(x) - \hat{x}\| \le \beta_p.\|x - \hat{x}\|^p$$

then $(x_n) = (F(x_{n-1}))$ converges of order p.

Proof :

Let $r > 0, B(\hat{x}, r) \subset U$, such that $\alpha = \beta_p.r^{p-1} < 1, \beta = \beta_p^{\frac{-1}{p-1}}$, $\gamma = -\frac{\log\alpha}{p(p-1)} > 0$. Let $x_1 \in B(\hat{x}, r)$, then

$$||x_1 - \hat{x}|| \leq r = \left[\frac{\alpha}{\beta_p}\right]^{\frac{1}{p-1}} = \beta.\exp\frac{\log\alpha}{p-1} = \beta.\exp(-\gamma p).$$

By induction we obtain from

$$||x_n - \hat{x}|| \leq \beta.\exp(-\gamma p^n)$$

the estimate

$$\begin{aligned}
||x_{n+1} - \hat{x}|| &\leq \beta_p||x_n - \hat{x}||^p \leq \beta_p\beta^p.\exp(-\gamma p^{n+1}) \\
&= \beta.\exp(-\gamma p^{n+1}) \leq r.
\end{aligned}$$

$\square$

Example :

If $f : [a, b] \to \mathcal{R}$ is p - times continuously differentiable at $\hat{t} = f(\hat{t}) \in [a, b]$, such that $f'(\hat{t}) = 0, ..., f^{(p-1)}(\hat{t}) = 0$, then $t_n = f(t_{n-1})$ is convergent of order p, since by Taylor

$$|f(t) - \hat{t}| = |f(t) - f(\hat{t})| = \frac{1}{p!}.|t - \hat{t}|^p.\sup_\tau |f^{(p)}(\tau)|.$$

If $g : [a, b] \to \mathcal{R}$ is twice continuously differentiable, $g(\hat{t}) = 0, g'(\hat{t}) \neq 0$, then

$$f(t) = t - \frac{g(t)}{g'(t)}$$

$$t_{n+1} = t_n - \frac{g(t_n)}{g'(t_n)} \qquad (3.1)$$

is convergent of order 2, since

$$\begin{aligned}
f'(t) &= 1 - \frac{g'(t)^2 - g''(t)g(t)}{g'(t)^2} \\
f'(t) &= 0.
\end{aligned}$$

Newton's Method

The equation (3.1) defines Newton's method for differentiable functions. Now we will consider Newton's method for solving the equation

$$F(x) = 0$$

in Banach spaces.

Theorem 3.1 *Let X, Y be Banach spaces and $F : B(x_0, r) \to Y$ continuously differentiable, such that*

(a) $F'(x_0)^{-1} \in \mathcal{L}(Y, X), \|F'(x_0)^{-1}.F(x_0)\| = \alpha,$
  $\|F'(x_0)^{-1}\| = \beta$
(b) $\|F'(u) - F'(v)\| \leq k\|u - v\|, \quad u, v \in B(x_0, r)$
(c) $2k\alpha\beta < 1, \quad 2\alpha < r$

are satisfied. Then F has a unique zero $\hat{x}$ in $\overline{B}(x_0, 2\alpha)$ and the Newton iterates

$$x_{n+1} = x_n - F'(x_n)^{-1}F(x_n) \tag{3.2}$$

converge quadratically to $\hat{x}$ and satisfy

$$\|x_n - \hat{x}\| \leq \frac{\alpha}{2^{n-1}}.q^{2n-1} = \frac{\alpha}{q.2^{n-1}}.exp(-\gamma 2^n) \tag{3.3}$$

with $q = 2\alpha\beta k < 1, \quad \gamma = -log \ q > 0.$

Proof :

At first we remark that for the real valued function $\omega(t) = y^*(F(u + t(v - u)))$. We obtain $(y^* \in Y^*)$

$$\omega'(t) = y^*(F'(u + t(v - u))(v - u))$$

$$\int_0^1 \omega'(t)dt = \omega(1) - \omega(0),$$

and therefore

$$F(v) - F(u) = \int_0^1 F'(u + t(v - u))(v - u)dt.$$

Suppose, that (x_n) is defined and let

$$\alpha_n = ||x_{n+1} - x_n||, \beta_n = ||F'(x_n)^{-1}||, \gamma_n = k\alpha_n\beta_n.$$

Then

$$
\begin{aligned}
\alpha_n &= ||F'(x_n)^{-1}F(x_n)|| \\
&\leq ||F'(x_n)^{-1}|| \ ||F(x_n) - (F(x_{n-1}) + F'(x_{n-1})(x_n - x_{n-1}))|| \\
&\leq \beta_n \int ||F'(x_{n-1} + t(x_n - x_{n-1})) - F'(x_{n-1})||dt. \\
&\leq ||x_n - x_{n-1}||\beta_n\alpha_{n-1}\int_0^1 k||t(x_n - x_{n-1})||dt \\
&\leq \beta_n\alpha_{n-1}^2 \cdot \frac{k}{2}.
\end{aligned}
$$

Since

$$
\begin{aligned}
F'(x_n) &= F'(x_{n-1}) + F'(x_n) - F'(x_{n-1}) \\
&= F'(x_{n-1}).[I + F'(x_{n-1})^{-1}(F'(x_n) - F'(x_{n-1}))],
\end{aligned}
$$

we also have

$$\beta_n = ||F'(x_n)^{-1}|| = ||(I+F'(x_{n-1})^{-1}(F'(x_n)-F'(x_{n-1})))^{-1}||.\beta_{n-1}$$

if

$$||F'(x_{n-1})^{-1}(F'(x_n) - F'(x_{n-1}))|| \leq \beta_{n-1}.k\alpha_{n-1} = \gamma_{n-1} < 1$$

then

$$\beta_n \leq \beta_{n-1}.(1 - \gamma_{n-1})^{-1},$$

hence

$$\alpha_n \leq \frac{k}{2}.\alpha_{n-1}^2.\beta_{n-1}(1 - \gamma_{n-1})^{-1}$$

and

$$\left.\begin{aligned}
\alpha_n &\leq \tfrac{1}{2}.\tfrac{\gamma_{n-1}}{1-\gamma_{n-1}}.\alpha_{n-1} \\
k\alpha_n\beta_n = \gamma_n &\leq \tfrac{1}{2}.\tfrac{\gamma_{n-1}^2}{(1-\gamma_{n-1})^2}
\end{aligned}\right\} \tag{3.4}$$

Newton's Method

Since $\gamma_0 = k\alpha_0\beta_0 \leq k\alpha\beta < \frac{1}{2}$, the inequalities (3.4) imply $\gamma_n < \frac{1}{2}$ and consequently $\alpha_n \leq \frac{1}{2}.\alpha_{n-1}$ for all $n \geq 1$. Hence, $\alpha_n \leq 2^{-n}\alpha$, and

$$||x_{n+1} - x_0|| \leq \sum_{j=0}^{n} ||x_{j+1} - x_j|| \leq \sum_{j=0}^{n} 2^{-j}\alpha \leq 2\alpha < r.$$

Thus it is obvious that (x_n) is well defined, and is a Cauchy sequence with $\lim x_n = \hat{x} \in \overline{B}(x_0, 2\alpha) \subset B(x_0, r)$.

Clearly, $F(\hat{x}) = 0$. Further

$$
\begin{aligned}
||x_{n+1} - \hat{x}|| &= ||x_n - \hat{x} - F'(x_n)^{-1}(F(x_n) - F(\hat{x}))|| \\
&\leq \beta_n ||F'(x_n)(x_n - \hat{x}) - F(x_n) - F(\hat{x})|| \\
&\leq \beta_n \int_0^1 ||F'(x_n - t(x_n - \hat{x})) - F'(x_n)||dt.||x_n - \hat{x}|| \\
&\leq \beta_n ||x_n - \hat{x}|| \int_0^1 kt||x_n - \hat{x}||dt \leq \frac{k}{2}\beta_n ||x_n - \hat{x}||^2,
\end{aligned}
$$

hence

$$||x_{n+1} - \hat{x}|| \leq c||x_n - \hat{x}||^2, c = \frac{k}{2}\sup \beta_n < \infty$$

since $\beta_n \to ||F'(\hat{x})^{-1}||$.

If $\hat{z} \in \overline{B}(x_0, 2\alpha)$ is another zero, then

$$
\begin{aligned}
||\hat{x} - \hat{z}|| &\leq \beta ||F(\hat{x}) - F(\hat{z}) - F'(x_0)(\hat{x} - \hat{z})|| \\
&\leq \beta k ||\hat{x} - \hat{z}||.\int_0^1 ||\hat{x} + t(\hat{z} - \hat{x}) - x_0||dt \\
&\leq 2\alpha\beta k ||\hat{x} - \hat{z}||
\end{aligned}
$$

and there $\hat{x} = \hat{z}$. Finally, to obtain equation (3.3), let $\delta_n = \gamma_n(1 - \gamma_n)^{-1}$. Then inequalities (3.4) and $\gamma_n \leq \frac{1}{2}$ imply

$$\delta_n = \gamma_n(1 - \gamma_n)^{-1} \leq 2\gamma_n \leq \delta_{n-1}^2,$$

hence

$$\delta_n \leq \delta_0^{2^n}$$

and consequently

$$\alpha_n \leq \frac{1}{2}\delta_{n-1}\alpha_{n-1} \leq \frac{1}{2}.\delta_0^{2^{n-1}}.\alpha_{n-1} \leq \ldots \leq 2^{-n}.\delta_0^{2^{n-1}}\alpha_0$$
$$\leq 2^{-n}.\delta_0^{2^{n-1}}\alpha, \text{ by } (3.4)$$

Therefore

$$\|x_n - \hat{x}\| \leq \lim_i \|x_n - x_{n+i}\| \leq \sum_{j=n}^{\infty} \alpha_j$$
$$\leq \sum_{j=n}^{\infty} 2^{-j}.q^{2^j-1}\alpha$$
$$\leq 2.2^{-n}.q^{2^n-1}\alpha = \frac{\alpha}{2^{n-1}}.q^{2^n-1} \ .$$

$\square$

Newton's method is quadratically convergent, provided that it is possible to find x_0 and $r > 0$, such that $\|F''(x_0)^{-1}\|$ and $\|F'(x_0)^{-1}F(x_0)\|$ are sufficiently small and F' is Lipschitz continuous in $B(x_0, r)$. Certainly this is another difficult problem without a general recipe for the solution.

A simplified Newton's type algorithm guarantees only linear convergence:

Theorem 3.2 *Let X, Y be Banach spaces and $F : B(x_0, r) \to Y$ continuously differentiable such that*

(a) $F'(x_0)^{-1} \in \mathcal{L}(Y, X), \|F'(x_0)^{-1}\| = \beta$,
$\|F'(x_0)^{-1}F(x_0)\| = \alpha$

(b) $\|F'(u) - F'(v)\| \leq k\|u - v\|, \quad u, v \in B(x_0, r)$

(c) $\gamma = 2k\alpha\beta < 1, r_0 = 2\alpha.\frac{1-\sqrt{1-\gamma}}{\gamma} \leq r$.

Then F has a unique zero $\hat{x} \in B(x_0, r_0)$ and the iterates of the modified Newton method

$$x_{n+1} = x_n - F'(x_0)^{-1}F(x_n)$$

The Implicit Function Theorem

converge to $\hat{x}$ and satisfy

$$||x_n - \hat{x}|| \leq \frac{q^n}{1-q}||x_1 - x_0||$$

with

$$q = 1 - \sqrt{1-\gamma}.$$

Proof :

The mapping $G : B(x_0, r_0) \to B(x_0, r_0)$, defined by

$$G(x) = x - F'(x_0)^{-1}F(x)$$

is a contraction. Let $x, y \in B(x_0, r_0)$

$$\begin{aligned}
G(x) - G(y) &= x - y - F'(x_0)^{-1}(F(x) - F(y)) \\
&= F'(x_0)^{-1}[F'(x_0)(x-y) - F(x) + F(y)] \\
&= F'(x_0)^{-1}\int_0^1 (F'(x_0) \\
&\quad -F'(y + t(x-y)))(x-y)dt.
\end{aligned}$$

$$\begin{aligned}
||G(x) - G(y)|| &\leq \beta k.||x_0 - y - t(x-y)||.||x-y|| \\
&\leq \beta k r_0 ||x-y||.
\end{aligned}$$

Since

$$r_0 = \frac{1 - \sqrt{1-\gamma}}{\beta k} < \frac{1}{\beta k} \ ,$$

G is a q-contraction with $q = \beta k r_0 = 1 - \sqrt{1-\gamma}$. It remains to show that G maps $B(x_0, r_0)$ into itself.

Let $x \in B(x_0, r_0)$. Then

$$||G(x) - x_0|| = ||x - x_0 - F'(x_0)^{-1}F(x)||$$

$$\leq \beta.||F'(x_0)(x - x_0) - F(x) + F(x_0)|| + \alpha$$

$$\leq \beta.\int_0^1 ||F'(x_0) - F'(x_0 + t(x - x_0))||dt.||x - x_0|| + \alpha$$

$$\frac{1}{2}.\beta k||x - x_0||^2 + \alpha \leq \frac{1}{2}\beta k r_0^2 + \alpha = r_0.$$

$\square$

Chapter 4

THE IMPLICIT FUNCTION THEOREM

Let X, Y, Z be Banach spaces and $U \subset X, V \subset Y$ neighbourhoods of x_0 and y_0 respectively. Let $F : U \times V \to Z$. In this chapter, we will consider the following problem.

If the equation $F(x, y) = 0$ has solution $x \in U$, if $y \in V$, provided that $F(x_0, y_0) = 0$. This is the generalization of the solution of $G(x) - y = 0$. The answer is given in the following theorems.

Theorem 4.1 *Let X, Y, Z be Banach spaces, $U \subset X, V \subset Y$ neighbourhoods of $x_0 \in U$, $y_0 \in V$. Let $F : U \times V \to Z$ be continuous and continuously differentiable with respect to y. Suppose that $F(x_0, y_0) = 0$ and $F_y^{-1}(x_0, y_0) \in \mathcal{L}(Z, Y)$. Then there exist balls $\overline{B}(x_0, r) \subset U, \overline{B}(y_0, s) \subset V$ and exactly one map $G : B(x_0, r) \to B(y_0, s)$, such that $G(x_0) = y_0$ and for all $x \in B(x_0, r)$*

$$F(x, G(x)) = 0.$$

This map G is continuous.

47

Proof :

Without loss of generality $x_0 = 0, y_0 = 0$ since the general situation may be reduced to this one by a translation.

Let $T = F'_y(0,0)$ and I be the identity on Y. Since $F(x,y) = 0$ is equivalent to

$$y + T^{-1}F(x,y) - y = 0,$$

we show that for $x \in U$

$$H(x,y) = T^{-1}F(x,y) - y$$

is a contraction on a properly chosen ball. Since $H'_y(0,0) = T^{-1}F'(0,0) - I = 0$ and H'_y is continuous, we can fix $k < 1$ and find $s > 0$, such that $||H'_y(x,y)|| \le k$ for $||x|| \le s, ||y|| \le s$. Thus,

$$||H(x,y_1) - H(x,y_2)|| \le k||y_1 - y_2||$$

by the mean value theorem.

Furthermore, since $H(0,0) = 0$, and $H(.,0)$ is continuous, there exist $r \le s$, such that

$$||H(x,0)|| < s(1-k), \quad \text{if } ||x|| \le r.$$

Then the mapping $\quad Q : B(0,s) \to Y$

$$Q(y) = -H(x,y)$$

maps $B(0,s)$ into itself and is a contraction:

$$\begin{aligned}
||Q(y)|| &\le ||H(x,y) - H(x,0)|| + ||H(x,0)|| \\
&\le k||y|| + k(1-s) < s.
\end{aligned}$$

Thus Q has a unique fixed point $y \in B(0,s)$, i.e. if $x \in B(0,r)$, then there is a unique $y = G(x)$, which is the fixed point of Q, hence

$$Q(y) = -H(x,G(x)) = y - T^{-1}F(x,G(x)) = y$$

The Implicit Function Theorem

this shows

$$F(x, G(x)) = 0.$$

If $x = 0$, then $y = 0$, since $0 - H(0,0) = 0$ implies that $H(0,y)$ has the fixed point $y = 0$, hence $G(0) = 0$. G is continuous, since

$$G(x_1) + H(x_1, G(x_1)) - G(x_2) - H(x_2, G(x_2)) = 0$$

implies

$$
\begin{aligned}
\|G(x_1) - G(x_2)\| &= \|H(x_1, G(x_1)) - H(x_2, G(x_2))\| \\
&\leq \|H(x_1, G(x_1)) - H(x_1, G(x_2))\| \\
&+ \|H(x_1, G(x_2)) - H(x_2, G(x_2))\| \\
&\leq k\|G(x_1) - G(x_2)\| \\
&+ \|H(x_1, G(x_2)) - H(x_2, G(x_2))\|
\end{aligned}
$$

hence

$$\|G(x_1) - G(x_2)\| \leq \frac{1}{1-k} \cdot \|H(x_1, G(x_2)) - H(x_2, G(x_2))\|$$

which tends to zero, if $x_1 \to x_2$.

$\square$

Corollary 4.1 *Let $F : U \to Y$ be continuous, $x_0 \in U$. If $F'(x_0)$ is continuously invertible, then there exist neighbourhoods U_0 of x_0 and V_0 of $y_0 = F(x_0)$, such that $F|_{U_0} = F_0$ is a homeomorphism and $G_0 = F_0^{-1}$ is differentiable at y_0*

$$G_0'(y_0) = F_0'(x_0)^{-1}.$$

Proof :

Let $\phi : Y \times U \to y$ be defined by

$$\phi(y, x) = F(x) - y.$$

Then $\phi_x'(y_0, x_0) = F'(x_0)$ is invertible. By the implicit function theorem there exist $U_0 \subset U, V_0 \subset Y$ and $G_0 : V_0 \to U_0$, such that for $y \in V_0$ we have

$$\phi(y, G_0(y)) = F(G_0(y)) - y = 0.$$

F is invertible, this implies by theorem 2.2, that G_0 is differentiable at y_0 with $G_0'(y_0) = F_0'(x_0)^{-1}$.

$\square$

Corollary 4.2 *Under the assumptions of Theorem 4.1 and additionally, that $F : U \times V \to Z$ is continuously differentiable (in both variables), then the map G is continuously differentiable and*

$$G'(x) = -F_y'(x, G(x))^{-1} F_x'(x, G(x)).$$

Proof :

Let $x, x + s \in B(0, r), t = G(x + s) - G(x)$, then $F(x + s, G(x) + t) = 0$. Then the differentiability guarantees for $\epsilon > 0$ the existence of $\delta > 0$, such that $||s|| \leq \delta$ implies

$$||F(x + s, G(x) + t) - F(x, G(x)) - F_x'(x, G(x))s - F_y'(x, G(x))t||$$

$$\leq \epsilon(||s|| + ||t||)$$

i.e.

$$||F_x'(x, G(x))x + F_y'(x, G(x))t|| \leq \epsilon(||s|| + ||t||).$$

Since $F_y'(0, 0)$ is an isomorphism, there exist a neighbourhood, such that $F_y'(x, G(x))$ is an isomorphism, thus

$$||t + F_y'(x, G(x))^{-1} F_x'(x, G(x))s|| \leq \epsilon^*(||s|| + ||t||). \qquad (*)$$

The Implicit Function Theorem

The definition of $t = G(x + s) - G(x)$ shows, that G is differentiable with derivative

$$G'(x) = -F_y'(x, G(x))^{-1} F_x'(x, G(x)),$$

if we show, that $\|t\| \leq \gamma \|s\|$. But this follows from the continuity of G and the inequality $(*)$.

$\square$

The solution of the implicit problem

$$\left. \begin{array}{c} F(x, y) = 0 \\ F(x_0, y_0) = 0 \end{array} \right\} \tag{4.1}$$

can be obtained by iterative methods of Banach or of Newton type.

Theorem 4.2 *Let $F : U \times V \to Z$ be as in Theorem 4.1. Let F_y' be continuous in (x_0, y_0). Let*

$$y_{n+1}(x) = y_n(x) - F_y'(x, y_0)^{-1} F(x, y_n(x)).$$

Then (y_n) converges to the solution G of equation (4.1), such that

$$\|y_n(x) - G(x)\| \leq q^n \|y_0(x) - G(x)\|$$

and

$$\begin{aligned} \|y_n(x) - G(x)\| &\leq \frac{q^n}{1 - q} \cdot \|y_1(x) - y_0(x)\| \\ &\leq \frac{q^n}{1 - q} \cdot \frac{\|T^{-1}\|}{1 - \epsilon \|T^{-1}\|} \cdot \|F(x, y_0(x))\| \end{aligned}$$

where ϵ, q are given by equations (4.2), (4.3) below.

Proof :

Let $T = F'_y(x_0, y_0)$ and let $p(r, s)$ be reals, such that for $x \in \overline{B}(x_0, r)$, $y_1, y_2 \in \overline{B}(x_0, s)$ by the mean value theorem and the continuity of F'

$$\|F(x_1, y_1) - F(x, y_2) - F'_y(x_0, y_0)(y_1 - y_2)\| \leq p(r, s)\|y_1 - y_2\|$$

with

$$\lim_{r, s \to 0} \|T^{-1}\| p(r, s) = 0.$$

If r_0, s_0 are sufficiently small, such that

$$\epsilon < \|T^{-1}\|^{-1} \tag{4.2}$$

$$q = \frac{\|T^{-1}\|}{1 - \epsilon\|T^{-1}\|}(p(r_0, s_0) + \epsilon) < 1 \tag{4.3}$$

$$\|F'_y(x, y_0) - F'_y(x_0, y_0)\| < \epsilon \tag{4.4}$$

$$\|F(x, y_0)\| \leq \frac{(1 - q)s_0(1 - \epsilon\|T^{-1}\|)}{\|T^{-1}\|} \tag{4.5}$$

then

$$H(x, y) = y - F'_y(x, y_0)^{-1}F(x, y)$$

has the property

$$\begin{aligned}
&\|H(x, y_1) - H(x, y_2)\| \\
= {}&\|y_1 - y_2 - F'_y(x, y_0)^{-1}.[F(x, y_1) - F(x, y_2)]\| \\
\leq {}&\|F'_y(x, y_0)^{-1}\|.\|F(x, y_1) - F(x, y_2) \\
&- F'_y(x, y_0)(y_1 - y_2)\|.
\end{aligned}$$

Since

$$\begin{aligned}
F'_y(x, y_0) &= F'_y(x_0, y_0) + F'_y(x, y_0) - F'_y(x_0, y_0) \\
&= T(I + T^{-1}(F'_y(x, y_0) - F'_y(x_0, y_0)))
\end{aligned}$$

The Implicit Function Theorem

we obtain

$$\|F_y'(x, y_0)^{-1}\| \le \frac{\|T^{-1}\|}{1 - \epsilon\|T^{-1}\|}$$

and

$$\begin{aligned}
\|H(x, y_1) - H(x, y_2)\| &\le \frac{\|T^{-1}\|}{1 - \epsilon\|T^{-1}\|} \cdot (p(r_0, s_0) + \epsilon) \cdot \|y_1 - y_2\| \\
&\le q \cdot \|y_1 - y_2\|.
\end{aligned}$$

Further

$$\begin{aligned}
\|H(x, y) - y_0)\| &\le \|H(x, y_0) - H(x, y_0)\| + \|H(x, y_0) - y_0\| \\
&\le q\|y - y_0\| + \|F_y'(x, y_0)^{-1}F(x, y_0)\| \\
&\le q \cdot s_0 + \frac{\|T^{-1}\|}{1 - \epsilon\|T^{-1}\|} \cdot \|F(x, y_0)\| \\
&\le q \cdot s_0 + \frac{\|T^{-1}\|}{1 - \epsilon\|T^{-1}\|} \cdot \frac{(1 - q)s_0(1 - \epsilon\|T^{-1}\|)}{\|T^{-1}\|} \\
&\le s_0.
\end{aligned}$$

By Banach's fixed point theorem we obtain the convergence and the error estimates.

$\square$

If we use Newton's method, then we will obtain superlinear convergence.

Theorem 4.3 *Let F be given as in Theorem 4.1. In a neighbourhood of (x_0, y_0) we assume the Hoelder condition*

$$\|F_y'(x, y_1) - F_y'(x, y_2)\| \le \gamma\|y_1 - y_2\|^\alpha$$

with $0 < \alpha \le 1$. Let F and F' be continuous at (x_0, y_0). Then there exist r_0, s_0, such that for $x \in B(x_0, r_0)$

$$y_{n+1} = y_n(x) - F_y'(x, y_n(x))^{-1}F(x, y_n(x))$$

Nonlinear Functional Analysis

exist and converge to the solution F of equation (4.1). Furthermore there exists a constant k_0 such that for any k with

$$k > [||T^{-1}||\gamma(1+\alpha)^{-1}]^{1/\alpha} = k_0$$

we obtain

$$||y_n(x) - G(x)|| \leq \frac{1}{2}[k||y_0(x) - G(x)||]^{(1+\alpha)^n}.$$

Proof :

We assume, that r_0, s_0 are small, that $F_y'(x,y)$ is continuously invertible whenever $x \in B(x_0, r_0)$, $y \in B(y_0, s_0)$ and $||F_y'(x,y)^{-1}|| \leq (1+\epsilon)||T^{-1}||$. Then

$$\begin{aligned}
&||y_n(x) - G(x)|| \\
=\ &||y_{n-1}(x) - G(x) - F_y'(x, y_{n-1}(x))^{-1} \\
&(F(x, y_{n-1}(x)) - F(x, G(x)))|| \\
\leq\ &(1+\epsilon)||T^{-1}||.||\int_0^1 F_y'(x, y_{n-1}(x)) - F_y'(x, G(x) \\
&+\tau(G(x) - y_{n-1}(x))d\tau||.||y_{n-1}(x) - G(x)|| \\
\leq\ &(1+\epsilon).||T^{-1}||.\gamma \int_0^1 (1-\tau)^\alpha d\tau.||y_{n-1}(x) - G(x)||^{1+\alpha} \\
\leq\ &\frac{(1+\epsilon)||T^{-1}||\gamma}{1+\alpha}.||y_{n-1}(x) - G(x)||^{1+\alpha},
\end{aligned}$$

hence

$$k||y_n(x) - G(x)|| \leq [k.||y_{n-1}(x) - G(x)||]^{1+\alpha}$$

where

$$k = \left[\frac{(1+\epsilon)||T^{-1}||.\gamma}{1+\alpha}\right]^{1/\alpha}.$$

We remark, that $k||G(x) - y_0||$ may be arbitrarily small, if the initial value $y_0(x)$ is sufficiently close to $G(x)$.

$\square$

Chapter 5

FIXED POINT THEOREMS

5.1 The Brouwer Fixed Point Theorem

Brouwer's fixed point theorem is basic for many fixed point theorems. It states that a continuous map, which maps a convex bounded closed set in $\mathcal{R}^n$ into itself, has a fixed point.

Before we prove the Brouwer fixed point theorem, we observe that the case of complex scalars is a consequence of the case of real scalars. This follows from the fact that the complex space $\mathcal{C}^n$ is isometric with the natural space $\mathcal{R}^{2n}$, and the unit spheres in these spaces correspond in a natural way. Thus we restrict our attention to real Euclidean space. We need following lemma.

Lemma 5.1 *Let f be an infinitely differentiable function of $n + 1$ variables $(x_0, ..., x_n)$ with values in $\mathcal{R}^n$. Let D_i denote the determinant whose colunmns are the n partial derivatives*

55

$f_{x_0}, ..., f_{x_{i-1}}, f_{x_{i+1}}, ..., f_{x_n}$. Then

$$\sum_{i=0}^{n}(-1)^i \cdot \frac{\partial}{\partial x_i} D_i = 0. \qquad (5.1)$$

Proof :

For every pair i, j of unequal integers between 0 and n, let C_{ij} denote the determinant whose first column is $f_{x_i x_j}$ and whose remaining columns are $f_{x_0}, .., f_{x_n}$ arranged in order of increasing indices, and where f_{x_i} and f_{x_j} are omitted from the enumeration. Clearly $C_{ij} = C_{ji}$, and by the laws governing differentiation of determinants and interchange of columns in them we have

$$\frac{\partial}{\partial x_i} D_i = \sum_{j<i}(-1)^j C_{ij} + \sum_{j>i}(-1)^{j-1} C_{ij}.$$

Hence

$$(-1)^i \cdot \frac{\partial}{\partial x_i} D_i = \sum_{i=0}^{n}(-1)^{i+j} C_{ij}\sigma(i,j),$$

where $\sigma(i,j) = 1$ if $j < i, \sigma(i,j) = 0$ if $i = j$, and $\sigma(i,j) = -1$ if $j > i$. Thus

$$\sum_{i=0}^{n}(-1)^i \cdot \frac{\partial}{\partial x_i} D_i = \sum_{i,j=0}^{n}(-1)^{i+j} C_{ij}\sigma(i,j).$$

Interchanging the dummy indices i, j in this latter expression and using the fact that $\sigma(i,j) = -\sigma(j,i)$, we see that

$$\sum_{i,j=0}^{n}(-1)^{i+j} C_{ij}\sigma(i,j) = \sum_{i,j=0}^{n}(-1)^{j+i} C_{ij}\sigma(j,i)$$

$$= (-1)\sum_{i,j=0}^{n}(-1)^{i+j} C_{ij}\sigma(i,j).$$

Thus all the three equal quantities in this formula must be zero, and formula (5.1) is proved.

$\square$

Fixed Point Theorems

Theorem 5.1 *(Brouwer)*

If ϕ is a continuous mapping of the closed unit sphere $B = \{x \in X, |x| \leq 1\}$ of Euclidean n-space into itself, then there is a point y in B such that $\phi(y) = y$.

Proof :

We have remarked that if suffices to consider real Euclidean space. Further, the Weierstrass approximation theorem for continuous functions of n variables implies that every continuous map ϕ of B into itself is the uniform limit of a sequence (ϕ_k) of infinitely differentiable mappings of B into itself. Suppose that the theorem were proved for infinitely differentiable maps. Then, for each integer k there is a point $y_k \in B$ such that $\phi_k(y_k) = y_k$. Since B is compact, some subsequences (y_{k_i}) converge to a point y in B. Since $\lim_{i \to \infty} \phi_{k_i}(x) = \phi(x)$ uniformly on B,

$$\phi(y) = \lim_{i \to \infty} \phi_{k_i}(y_{k_i}) = \lim_{i \to \infty} y_{k_i} = y.$$

This shows that it is sufficient to consider the case that ϕ is infinitely differentiable.

We suppose that ϕ is an infinitely differentiable map of B into itself and that $\phi(x) \neq x, x \in B$. Let $a = a(x)$ be the larger root of the quadratic equation $|x + a(x - \phi(x))|^2 = 1$, so that

$$\begin{aligned}
1 &= (x + a(x - \phi(x)), x + a(x - \phi(x))) \\
&= |x|^2 + 2a(x, x - \phi(x)) + a^2|x - \phi(x)|^2.
\end{aligned}$$

By the quadratic formula

$$\begin{aligned}
a(x).|x - \phi(x)|^2 = \ &(x, \phi(x) - x) \\
&+ \{(x, x - \phi(x))^2 \\
&+ (1 - |x|^2)|x - \phi(x)|^2\}^{\frac{1}{2}}.
\end{aligned} \qquad (5.2)$$

58   *Nonlinear Functional Analysis*

Since $|x - \phi(x)| \neq 0$ for $x \in B$, the discriminant $(x, x - \phi(x))^2 + (1 - |x|^2)|x - \phi(x)|^2$ is positive when $|x| \neq 1$. Also if $|x| = 1$, then $(x, x - \phi(x)) \neq 0$, for otherwise $(x, \phi(x)) = 1$ and the inner product of two vectors with length at most 1 can be equal to 1 only when they are equal. Thus the discriminant is never zero for x in B. Since the function $t^{1/2}$ is an infinitely differentiable function of t for $t > 0$, and since $|x - \phi(x)| \neq 0, x \in B$ it follows from formula (5.2) that the function $a(x) = 0$ for $|x| = 1$ is an infinitely differentiable function of $x \in B$. Moreover, it follows from formula (5.2), that $a(x) = 0$ for $|x| = 1$. Now, for each real number t, put $f(t; x) = x + ta(x)(x - \phi(x))$. Then f is an infinitely differentiable function of the $n + 1$ variables $t, x_1, ..., x_n$ with values in B. Since $a(x) = 0$ for $|x| = 1$, we have $f_t'(t, x) = 0$ for $|x| = 1$. Also $f(0, x) = x$, and from the definition of a we have $|f(1, x)| = 1$ for all $x \in B$.

Denote the determinant whose columns are the vector $f_{x_1}'(t, x)$, ..., $f_{x_n}'(t, x)$ by $D_0(t, x)$ and consider the integral

$$I(t) = \int_B D_0(t, x)dx. \qquad (5.3)$$

It is clear that $I(0)$ is the volume of B and hence $I(0) \neq 0$. Since $f(1, x)$ satisfies the functional dependence $|f(1, x)| = 1$. It follows that the Jacobian determinant $D_0(1, x)$ is identically zero, hence $I(1) = 0$. The desired contradiction will be obtained if we can show that $I(t)$ is a constant; i.e., that $I'(t) = 0$. To prove this, differentiate under the integral sign and employ (5.1) to conclude that $I'(t)$ is a sum of integrals of the form

$$\pm \int_B \frac{\partial}{\partial x_i} D_i(t, x)dx$$

where $D_i(t, x)$ is the determinant whose columns are the vectors

$$f_t'(t, x), f_{x_1}'(t, x), ..., f_{x_{i-1}}'(t, x), f_{x_{i+1}}'(t, x), ..., f_{x_n}'(t, x).$$

Fixed Point Theorems

By the *Gauß* Theorem we see that

$$\int_B \frac{\partial}{\partial x_i} D_i(t,x)dx = \int_{\partial B} D_i(t,x)\eta_i d\omega.$$

For $x \in \partial B, |x| = 1$, and

$$f'_t(t,x) = a(x)(x - \phi(x))$$

implies $f'_t(t,x) = 0$ for $x \in \partial B$, hence $D_i(t,x) = 0$. This implies $I'(t) = 0, I = $ constant, $I(1) = 0 \neq I(0)$. This contradiction shows, that there is an $x \in B$ such that $\phi(x) = x$.

$\square$

A simple application of the Brouwer fixed point theorem is the following existence principle for systems of equations.

Proposition 5.1 *Let $B = \overline{B}(0,r) \subset \mathcal{R}^n$ be the closed ball with radius r and $g_j : B \to \mathcal{R}$ be continuous mappings, $j = 1, 2, ..., n$. If for all $x = (\xi_1, \xi_2, ..., \xi_n) \in \mathcal{R}^n, ||x|| = r$*

$$\sum_{j=1}^n g_j(x)\xi_j \geq 0 \tag{5.4}$$

then the system of equations

$$g_j(x) = 0, \quad j = 1, 2, ..., n \tag{5.5}$$

has a solution $\hat{x}$ with $||\hat{x}|| \leq r$.

Proof :

Let $g(x) = (g_1(x), ..., g_n(x))$ and assume $g(x) \neq 0$ for all $x \in B$. Define

$$f(x) = -\frac{rg(r)}{||g(x)||}.$$

f is a continuous map of the compact convex set B into itself. Therfore there exists a fixed point $\tilde{x}$, of f with $||\tilde{x}|| = ||f(\tilde{x})|| = r$. Furthermore

$$\Sigma g_j(\tilde{x})\xi_j = -\frac{1}{r}||g(\tilde{x})||.\Sigma f_j(\tilde{x})\xi_j$$

$$= -\frac{1}{r}||g(\tilde{x})||.\Sigma \xi_j^2 < 0.$$

This contradicts formula (5.4), hence equation (5.5) has a zero.

$\square$

5.2 The Schauder Fixed Point Theorem

We will begin this section with a theorem on the extension of continuous mappings. Let $A \subset X$ be a subset of the normed space X.

A is **convex**, if for all $x, y \in A, \tau \in [0, 1]$, the segment $\tau x + (1 - \tau)y \in A$. Then by induction, if $x_1, ..., x_n \in A$, $\tau_i \geq 0, \sum_{i=1}^n \tau_i = 1$, then $\sum_{i=1}^n \tau_i x_i \in A$. The **convex hull** coB of $B \subset X$ is the intersection of all convex sets A, which contain B. It is given by

$$coB = \{x = \sum_{i=1}^n \tau_i x_i, x_i \in B, \tau_i \geq 0, \sum_{i=1}^n \tau_i = 1, n \in \mathcal{N}\}.$$

Theorem 5.2 *Let X be a Banach space, $A \subset X$ a closed subset and $F : A \to Y$ a continuous map from A into the Banach space Y. Then there exists a continuous extension $\tilde{F}$ of F with $\tilde{F} : X \to Y, \tilde{F}|_A = F$ and*

$$\tilde{F}(x) \subset coF(a).$$

Fixed Point Theorems

61

Proof :

(a) We construct a partition of unity.

Let $x \in X \setminus A, B_x = B(x, r_k)$ be an open ball with diameter $2r_x$ less than the distance of A to B_x, e.g.

$$r_x < \frac{1}{6} \ \text{dist}(x, A)$$

Then

$$X \setminus A = \cup_{x \in X \setminus A} B_x.$$

A result of General Topology states, that there exists a locally finite open refinement $\{U_\lambda, \lambda \in \Lambda\}$ of $\{B_x, x \in X \setminus A\}$, i.e. there exists a covering of $X \setminus A$, such that

$$\forall x \in X \setminus A \ \exists_{B_x} \ \varphi\{\lambda : U_\lambda \subset B_x\} < +\infty.$$

Let $\alpha : X \setminus A \to \mathcal{R}$

$$\alpha(x) = \sum_{\lambda \in \Lambda} \ \text{dist}(x, X \setminus U_\lambda).$$

This sum is always finite, and $x \in X \setminus \Rightarrow \alpha(x) > 0$. Let

$$\varphi_\lambda(x) = \frac{\text{dist}(x, X \setminus U_\lambda)}{\alpha(x)}.$$

φ_λ is continuous, $0 \leq \varphi_\lambda \leq 1, \Sigma \varphi_\lambda(x) = 1. \{\varphi_\lambda, \lambda \in \Lambda\}$ is a partition of unity for $X \setminus A$.

(b) We now construct the extension.

$\forall \lambda \in \Lambda \exists x \in X \setminus A \ U_\lambda \subset B_x$, hence

$$\text{dist}(A, U_\lambda) \geq \text{dist}(A, B_x) > 2r_x > 0.$$

For every $\lambda \in \Lambda$ we choose $a_\lambda \in A$, such that

$$\text{dist}(a_\lambda, U_\lambda) < 2\text{dist}(A, U_\lambda).$$

Let

$$\tilde{F}(x) = \begin{cases} F(x) & , \text{ if } x \in A \\ \Sigma\varphi_\lambda F(a_\lambda) & , \text{ if } x \in X \setminus A \end{cases}$$

Then $\tilde{F} : x \to Y$ is an extension of F, and $\tilde{F}(x) \subset co(F(A))$, since $\Sigma\varphi_\lambda(x) = 1, \varphi_\lambda(x) \geq 1$.

(c) $\tilde{F}$ is continuous :

$\tilde{F}$ is continuous for all $x \in \partial A$. Let $x_0 \notin \partial A$. Since F is continuous at x_0,

$$\forall \epsilon > 0 \ \exists \delta > 0 \ \forall x \in A \ ||x - x_0|| < \delta$$
$$\Rightarrow \ ||F(x) - F(x_0)|| < \epsilon.$$

This implies

$$\forall x \in X \ ||x - x_0|| \ < \ \delta/4 \Rightarrow ||\tilde{F}(x) - \tilde{F}(x_0)|| < \epsilon$$
$$\tilde{F}(x) - \tilde{F}(x_0) \ = \ \Sigma\varphi_\lambda(x)(F(a_\lambda) - F(x_0))$$
$$||\tilde{F}(x) - \tilde{F}(x_0)|| \ \leq \ \Sigma\varphi_\lambda(x)||F(a_\lambda) - F(x_0)||.$$

If $\varphi_\lambda(x) \neq 0$, then $x \in U_\lambda$ and

$$||x - a_\lambda|| \ \leq \ ||x - u|| + ||u - a_\lambda|| \leq \text{diam } U_\lambda + ||u - a_\lambda||$$
$$\leq \ 2r_x + \text{dist}(a_\lambda, U_\lambda)$$
$$\leq \ \text{dist}(A, B_x) + 2 \ \text{dist}(A, U_\lambda)$$
$$\leq \ 3 \ \text{dist}(A, B_x) \leq 3||x_0 - x||$$

and

$$||x_0 - a_\lambda|| \leq ||x_0 - x|| + ||x - a_\lambda|| \leq 4||x_0 - x|| < \delta$$

together with $||F(a_\lambda) - F(x_0)|| < \epsilon$ implies

$$||\tilde{F}(a_\lambda) - \tilde{F}(x_0)|| \leq \Sigma\varphi_\lambda(x).\epsilon = \epsilon.$$

Therefore $\tilde{F} : X \to Y$ is continuous.

$\square$

Fixed Point Theorems

Corollary 5.1 *Let $C \subset X$ be a closed convex subset of X. Then there exists a mapping $R : X \to C$, such that $R|_C = Id$.*

Proof :

For $F = Id|_c$ apply the extension theorem.

$\square$

Corollary 5.2 *(Brouwer's Fixed Point Theorem)*

Let C be a compact convex set in $\mathcal{R}^n$, $f : C \to C$ a continuous mapping. Then f has a fixed point.

Proof :

Let $r > 0$, such that $\overline{B}(0,r) \supset C$, let $\tilde{f} : \mathcal{R}^n \to \mathcal{R}^n$ be an extension of f with $\tilde{f}(\mathcal{R}^n) \subset coC$. Then

$$\tilde{f}(\overline{B}(0,r)) \subset coC \subset C \subset \overline{B}(0,r).$$

$\tilde{f}$ has a fixed point $\hat{x}$ in C, since $f(\hat{x}) = \hat{x} \in C$.

$\square$

Theorem 5.3 *(Schauder's Fixed Point Theorem)*

Let $K \subset X$ be a convex compact set, and $F : K \to K$ continuous. Then F has a fixed point.

Proof :

Let $\epsilon > 0$, and $x_1, ..., x_n$ an ϵ - net for K, i.e.

$$K \subset U\{x_j + \epsilon B\}, \quad \text{or} \quad \forall x \in K \; \exists x_j \|x - x_j\| \leq \epsilon.$$

$(B = B(0,1) = \{x \in X : \|x\| \leq 1\})$.

Let

$$h_j(x) = \begin{cases} 0, & \text{if } \|x - x_j\| > \epsilon \\ \epsilon - \|x - x_j\|, & \text{if } \|x - x_j\| \leq \epsilon \end{cases}$$

64   *Nonlinear Functional Analysis*

and

$$h_\epsilon(x) = \frac{\Sigma h_j(x) x_j}{\Sigma h_j(x)}.$$

h is continuous, if $x \in K$, then

$$||h(x) - x|| = ||\frac{\Sigma h_j(x)(x_j - x)}{\Sigma h_j(x)}|| \leq \epsilon.$$

Let $K_0 = \overline{co}\{x_1, ..., x_n\}$. K_0 is compact, $K_0 \subset K$, and

$$h_\epsilon o F : K_0 \to K_0.$$

By Brouwer's fixed point theorem $h_\epsilon \ o \ F$ has a fixed point $x_\epsilon \in K_0$, hence

$$h_\epsilon(F(x_\epsilon)) = x_\epsilon$$

and

$$||F(x_\epsilon) - x_\epsilon|| = ||F(x_\epsilon) - h_\epsilon(F(x_\epsilon))|| \leq \epsilon.$$

The set $\{x_\epsilon, \epsilon > 0\} \subset K$ has a point of accumulation $\hat{x} \in K$, hence

$$||F(\hat{x}) - \hat{x}|| \leq ||F(\hat{x}) - F(x_\epsilon)|| + ||F(x_\epsilon) - x_\epsilon|| + ||x_\epsilon - \hat{x}||$$

$\lim_{h \to \infty} x_{\epsilon_n} = \hat{x}$ implies $F(\hat{x}) = \hat{x}$.

$\square$

Chapter 6

SET CONTRACTIONS AND DARBO'S FIXED POINT THEOREM

6.1 Measures of Noncompactness

Let $\mathcal{B}$ be the family of all bounded subsets of the Banach space X; recall that $B \subset X$ is bounded if B is contained in some ball.

$B \in \mathcal{B}$ is relatively compact, if there exists for any $\epsilon > 0$ a finite covering of B by balls of radius ϵ. If $B \in \mathcal{B}$ is not relatively compact then there exists an $\epsilon > 0$, such that B cannot be covered by a finite number of ϵ - balls, and it is then impossible to cover B by finitely many sets of diameter $< \delta$. Recall that

$$\text{diam} \quad B = \sup\{\|x - y\|, x, y \in B\}$$

is called the **diameter** of B.

Definition 6.1 *Let X be a Banach space and $\mathcal{B}$ its bounded sets. Then $\chi : \mathcal{B} \to \mathcal{R}^+$, defined by*

$\chi(B) = inf\{\delta > 0, B$ *admits a finite cover by sets of diameter $\leq \delta\}$ is called the (Kuratowski-) measure of noncom-*

65

pactness and

$\beta : \mathcal{B} \to \mathcal{R}^+$, *defined by*

$\beta(B) = inf\{\rho > 0, B$ *can be covered by finitely many balls of radius ρ } is called the ball measure of noncompactness.*

Evidently, for $B \in \mathcal{B}$

$$\beta(B) \le \chi(B) \le 2\beta(B),$$

but there exist $B \in \mathcal{B}$, such that the strict inequalities hold.

The properties of the measures of noncompactness are collected in the following statements.

Proposition 6.1 *Let X be a Banach space, $B, B_j \in \mathcal{B}$. Then*

1^0 $\chi(\phi) = 0$

2^0 $\chi(B) = 0$ *iff B is relatively compact*

3^0 $0 \le \chi(B) \le diam B$

4^0 $B_1 \subset B_2 \Rightarrow \chi(B_1) \le \chi(B_2)$

5^0 $\chi(B_1 + B_2) \le \chi(B_1) + \chi(B_2)$

6^0 $\chi(\lambda B) = |\lambda|\chi(B) \quad \lambda \in \mathbf{K}$

7^0 $\chi(B) = \chi(\overline{B})$

8^0 $\chi(B_1 \cup b_2) = max(\chi(B_1), \chi(B_2)).$

Set Contractions and Darbo's Fixed Point Theorem 67

Proof :

1^0 By definition diam $\phi = 0$.

2^0 B is relatively compact iff for every $\epsilon > 0$ there exists a finite covering by balls of diameter ϵ.

3^0 M can be covered by M with diam M.

4^0 Every cover of B_2 is a cover of B_1.

5^0 Let $M_1, ..., M_m$ be a cover of $B_1, N_1, ..., N_n$ a cover of B_2, then all sets $M_j + N_k$ form a cover of $B_1 + B_2$ and
$$\text{diam}(M_j + n_k) \leq \text{ diam } m_j + \text{diam } N_k.$$

6^0 Note that diam $(\lambda B) = |\lambda|\text{diam } B$.

7^0 From $B \subset \overline{B}$ follows $\chi(B) \leq \chi(\overline{B})$. Conversely if $B \subset \cup M_j$, then $\overline{B} \subset \cup \overline{M}_j$ with diam $M_j = \text{diam}_j$, so $\chi(B) \leq \chi(\overline{B})$.

8^0 Let $B = B_1 \cup B_2$ and $\beta = \max\{\chi(B_1), \chi(B_2)\}$. Then it follows from $B_j \subset B$ that $\chi(B) \leq \chi(B)$ and $\beta \leq \chi(B)$.
Conversely let for $\epsilon > 0$ given convergins $B_j \subset M_{jk}$ with diam $M_{jk} \leq \chi(B_j) + \epsilon \leq \beta + \epsilon$. All of these M_{jk}'s together form a covering of B, so that $\chi(B) \leq \beta + \epsilon$, i.e. $\chi(B) \leq \beta$.

$\square$

Proposition 6.2 *Let B be a bounded set in X, then*

$$\chi(B) = \chi(coB) = \chi(\overline{co}B).$$

Proof :

Since $B \subset coB, \chi(B) \leq \chi(coB)$, it remains to be shown that $\chi(coB) \leq \chi(B)$. It is

$$\text{diam } B = \text{diam } coB,$$

since for $x, y \in coB$

$$x = \Sigma \lambda_j x_j \qquad x_j \in B, \lambda_j \geq 0, \Sigma \lambda_j = 1$$
$$y = \Sigma \mu_k y_k \qquad y_k \in B, \mu_k \geq 0, \Sigma \mu_k = 1$$

$$\begin{aligned} x - y &= \Sigma \lambda_j x_j - y = \Sigma \lambda_j (x_j - y) \\ &= \Sigma \lambda_j (x_j - \Sigma \mu_k y_k) \\ &= \Sigma \lambda_j \Sigma \mu_k . (x_j - y_k) \end{aligned}$$

$$\begin{aligned} \|x - y\| &\leq \Sigma \lambda_j \ \Sigma \mu_k . \ \|x_j - y_k\| \\ &\leq \Sigma \lambda_j \ \Sigma \mu_k \ \text{diam } B = \text{diam } B. \end{aligned}$$

Now, let $B \subset \cup_{j=1}^m M_j$ with diam $M_j \leq \chi(B) + \epsilon$; and we can assume, that the M_j are convex. Let

$$\Lambda = \{(\lambda_1, ..., \lambda_m) \in \mathcal{R}^m, \sum_{j=1}^m \lambda_j = 1, \lambda_j \geq 0\}$$

and for $\lambda \in \Lambda$

$$A(\lambda) = \sum_{j=1}^m \lambda_j M_j.$$

Then

$$\chi(A(\lambda)) \leq \Sigma \lambda_j \chi(M_j) \leq \chi(B) + \epsilon.$$

Now we show that

$$C = \cup_{\lambda \in \Lambda} A(\lambda) \text{ is convex.}$$

Set Contractions and Darbo's Fixed Point Theorem

Let $x, y \in C$, i.e. $x \in A(\lambda), y \in A(\mu)$, and $t \in [0,1]$. Then

$$x = \Sigma \lambda_j x_j, \quad \lambda \in \Lambda, \quad x_j \in M_j$$

$$y = \Sigma \mu_j y_j, \quad \mu \in \Lambda, \quad y_j \in M_j$$

Let $\nu_j = t\lambda_j + (1-t)\mu_j$,

$$\rho_j = \begin{cases} 0 & \text{if } \nu_j = 0 \\ t\lambda_j/\nu_j & \text{if } \nu_j \neq 0 \end{cases}$$

then $0 \leq \rho_j \leq 1, z_j = \rho_j x_j + (1-\rho_j)y_j \in M_j$ and

$$\begin{aligned}
tx + (1-t)y &= \Sigma t\lambda_j x_j + (1-t)\mu_j y_j \\
&= \Sigma \nu_j \rho_j x_j + (\nu_j - t\lambda_j)y_j \\
&= \Sigma \nu_j \rho_j x_j + \nu_j(1-\rho_j)y_j \\
&= \Sigma \nu_j z_j \in A(\nu) \subset C
\end{aligned}$$

with $\nu = (\nu_1, ..., \nu_n) \in \Lambda$.

Finally we have to show the finiteness. Since

$$B \subset \cup_{j=1}^{n} M_j \subset \cup_{\lambda \in \Lambda} A(\lambda) = C, \quad coB \subset C.$$

Since Λ is compact, there exist $\lambda^{(1)}, ..., \lambda^{(r)} \in \Lambda$ such that for $\lambda \in \Lambda$ there is a $\lambda^{(k)}$, such that

$$\sum_{j=1}^{m} |\lambda_j - \lambda_j^{(k)}| \sup_{x \in B} \|x\| \leq \epsilon.$$

Let $x \in A(\lambda), x = \sum_{j=1}^{m} \lambda_j x_j$. Let $y = \Sigma \lambda_j^{(k)} x_j$. Then

$$\|x - y\| \leq \Sigma |\lambda_j - \lambda_j^{(k)}|.\|x_j\| \leq \epsilon.$$

This means (U denotes the unit ball)

$$C = \cup_{\lambda \in \Lambda} A(\lambda) \subset \cup_{k=1}^{r} A(\lambda^{(k)}) + \epsilon U$$

$$\chi(C) \leq \max_k \chi(A(\lambda^{(k)})) + \epsilon \leq \chi(B) + 2\epsilon.$$

Thus

$$\chi(coB) \leq \chi(C) \leq \chi(B) + 2\epsilon.$$

This implies the equality

$$\chi(coB) = \xi(B).$$

$\square$

Proposition 6.3 *Let X be a Banach space and (M_j) a decreasing sequence of nonempty closed bounded subsets of X, i.e.*

$$M_1 \supseteq M_2 \supseteq ... \supseteq M_n \supseteq ...$$

such that

$$\lim_{n \to \infty} \chi(M_n) = 0.$$

Then

$$M_\infty = \cap_{n \in \mathcal{N}} M_n$$

is nonempty and compact.

Proof :

M_∞ is compact, since $\chi(M_\infty) \leq \chi(M_m) \to 0$ and M_∞ is closed. We have to show that $M_\infty \neq \phi$. Let M_{kj} be chosen, such that

$$M_k = \cup_{j=1}^{j(k)} M_{kj}$$

$$\chi(M_{kj}) \leq \chi(M_k) + \frac{1}{k}.$$

Claim :

$$\exists j, \ \forall n \in \mathcal{N} \ M_{j1} \cap M_n \neq \phi.$$

If not, there is for each j an index n_j, such that

$$M_{nj} \cap M_{ij} = \phi,$$

Set Contractions and Darbo's Fixed Point Theorem

then, for all $n \geq n_j$

$$M_n \cap M_{1j} = \phi$$

therefore

$$M_n = M_n \cap M_1 = M_n \cap (\cup M_{1i}) \subset M_n \cap M_{1j} = \phi.$$

This contradicts $M_n \neq \phi$.

Claim :

$$\exists j_2 \ \forall n \in \mathcal{N} \ \ M_{1j_1} \cap M_{2j_2} \cap M_n \neq \phi.$$

If not, then for each j there is an index n_j, such that for $n \geq n_j$

$$M_{1j_1} \cap M_{2j_2} \cap M_n = \phi$$

which contradicts

$$M_n \subset M_2 \subset \cup_j M_{2j}$$

and $M_n \cap M_{1j_1} \neq \phi$ for all n.

Finally by induction, we find that for every k there is an index j_k such that for all n, m

$$M_n \cap (\cup_{k=1}^m M_{kj_k}) \neq \phi. \qquad (*)$$

Let

$$N_m = \cup_{k=1}^m M_{kj_k}$$

then $N_1 \supseteq N_2 \supseteq ... \supseteq N_m \supseteq ...$ and $M_m \subset M_{mj_m}$.

$(*)$ implies, that for all m

$$N_m \neq \phi$$

$$\text{diam} \ N_m \leq \ \text{diam} \ M_m \leq \chi(M_m) + \frac{1}{m}$$

hence

$$\lim \ \text{diam} \ N_m = 0.$$

72   *Nonlinear Functional Analysis*

Choose a sequence (z_m), such that $z_m \in N_m$. Then (z_m) is a Cauchy sequence, which is convergent,

$$\lim z_m = z \in \cap N_m \subset \cap M_n = M_\infty.$$

Therefore M_∞ is nonempty.

$\square$

6.2   Condensing Maps

A continuous map F is called **bounded**, if it maps bounded sets onto bounded sets, and **compact**, if it maps bounded sets into compact sets.

Definition 6.2 *Let X be a Banach space and $U \subset X$. An operator $F : U \to X$ is called a k-set contraction, $0 \leq k < 1$, iff F is continuous, maps bounded sets onto bounded sets, such that for all bounded sets $B \subset U$*

$$\chi(F(B)) \leq k\chi(B).$$

F is said to be condensing, if for all bounded sets B with $\chi(B) \neq 0$

$$\chi(F(B)) < \chi(B).$$

Obviously, every k-set contraction is condensing. Every compact map is a 0-set contraction. The following example is an important one.

Example 6.4

Let X be a Banach space, $U \subset X$ and $K : U \to X$ Lipschitz continuous with Lipschitz constant $k < 1$. $C : U \to X$ is compact.

Set Contractions and Darbo's Fixed Point Theorem 73

Then

$$F = K + C$$

is a k-set contraction.

Proof :

Let $B \subset U$ be a bounded set. Then

$$
\begin{aligned}
F(B) &\subset K(B) + C(B) \\
\chi(F(B)) &\leq \chi(K(B)) + \chi(C(B)) \\
&\leq k\chi(B).
\end{aligned}
$$

$\square$

A continuous map is said to be **proper**, if the preimage of each compact set is compact.

Lemma 6.1 *Let $B \subset X$ be closed and bounded and $F : B \to X$ condensing. Then $I - F$ is proper and $I - F$ maps closed subsets of B onto closed sets.*

Proof :

Let K be compact and $A = (I - F)^{-1}K$ the preimage of K under $I - F$. Since $I - F$ is continuous, A is closed. From $K = (I - F)A$ we see, $A \subset F(A) + K$, hence

$$\chi(A) \leq \chi(F(A)) + \chi(K) = \chi(F(A)).$$

Since F is condensing, $\chi(A) = 0$, i.e. A is compact.

Now let A be closed and $(x_n) \subset A, y_n = (I - F)x_n$. Let $y = \lim y_n$. The set $K = \{y\} \cup \{y_n, n \in \mathcal{N}\}$ is compact, i.e. $\lim x_n = x$ exists and belongs to A, since A is closed. Thus $x \in A$ implies

$$y = (I - F)x \in (I - F)A,$$

74   *Nonlinear Functional Analysis*

and $(I - F)A$ is closed.

$\square$

We are now able to prove the fixed point theorem of Darbo (1955) and Sadovski (1967) which is a common generalization of Banach's and of Schauder's fixed point theorem.

Theorem 6.1 *Let $C \subset X$ be a nonempty, closed, bounded and convex subset of a Banach space X and let $F : C \to C$ be condensing. Then F has a fixed point.*

Proof :

Without loss of generality we may assume, that $0 \in C$.

Let F be a strict k-set contraction with $k < 1$. Define the decreasing sequence $C_0 = C, C_1 = \overline{co}F(C_0), C_n = \overline{co}F(C_{n-1})$. We have

$$C_0 \supseteq C_1 \supseteq C_2 \supseteq \ldots \supseteq C_n \supseteq \ldots$$

$$\chi(C_n) \le k\chi(C_{n-1}) \le \ldots \le k^n\chi(C_0),$$

hence

$$\lim \chi(C_n) = 0$$

and

$$C_\infty = \cap_{n \in \mathcal{N}} C_n$$

is compact. Furthermore, C_∞ is convex and

$$F(C_\infty) \subset C_\infty.$$

Therefore, Schauder's fixed point theorem shows, that F has a fixed point in $C_\infty \subset C$.

Now let F be condensing and let (k_n) be a sequence with $0 \le k_n < 1$, $\lim k_n = 1$. Then, since $0 \in C, F_n = k_nF$ maps C into C, and has a fixed point x_n,

The Topological Degree

thus

$$x_n - F(x_n) = x_n - k_n F(x_n) - (1 - k_n)F(x_n)$$

implies

$$\lim x_n - F(x_n) = 0.$$

The set $K = \{0\} \cup \{x_n - F(x_n)\}$ is compact. Thus, since $I - F$ is proper by Lemma 6.5, $(I - F)^{-1}K$ is compact, i.e. $\{x_n, n \in \mathcal{N}\}$ has a point of accumulation $\hat{x}$, i.e. there exists a subsequence with $\lim x_{n_k} = \hat{x} = F(\hat{x}) = \lim F(x_{n_k})$. Therefore F has a fixed point $\hat{x}$ in C.

$\square$

Chapter 7

THE TOPOLOGICAL DEGREE

7.1 Axiomatic Definition of the Brouwer Degree in $\mathcal{R}^n$

The Brouwer degree is a tool, which allows an answer to the question, if a given equation

$$f(x) = y$$

has a solution $x \in \Omega$, where $\Omega \subset \mathcal{R}^n$ is open and bounded, and $f : \overline{\Omega} \to \mathcal{R}^n$ is continuous, and y does not belong to the image $f(\partial\Omega)$ of the boundary $\partial\Omega$ of Ω. More precisely, for each admissible triple (f, Ω, y) we associate an integer $d(f, \Omega, y)$ such that $d(f, \Omega, y) \neq 0$ implies the existence of a solution $x \in \Omega$ of this equation $f(x) = y$. This integer is uniquely defined by the following properties:

If f is the identity map, and $y \in \mathcal{R}^n$, then $f(x) = y$ has a solution $x \in \Omega$, iff $y \in \Omega$, i.e.

$$(d1) \qquad d(id, \Omega, y) = \begin{cases} 1 & \text{for } y \in \Omega \\ 0 & \text{for } y \notin \overline{\Omega} \end{cases}$$

77

Nonlinear Functional Analysis

The second condition is a natural formulation of the desire that d should yield information on the location of solutions. Suppose that Ω_1 and Ω_2 are disjoint open subsets of Ω and suppose, that $f(x) = y$ has finitely many solutions in $\Omega_1 \cup \Omega_2$, but no solution in $\overline{\Omega} \setminus (\Omega_1 \cup \Omega_2)$. Then the number of solutions in Ω is the sum of the numbers of solutions in Ω_1 and Ω_2, and this suggests that d should be additive in its argument Ω, that is

$$(d2) \qquad d(f, \Omega, y) = d(f, \Omega_1, y) + d(f, \Omega_2, y)$$

whenever Ω_1 and Ω_2 are disjoint open subsets of Ω, such that $y \notin f(\overline{\Omega} \setminus (\Omega_1 \cup \Omega_2))$.

The third and the last condition reflects the desire, that for complicated f the number $d(f, \Omega, y)$ can be calculated by $d(g, \Omega, y)$ with simpler g, atleast if f can be continuously deformed into g such that at no stage of the deformation we get solutions on the boundary. This leads to

$$(d3) \qquad d(h(t, .), \Omega, y(t)) \text{ is independent of } t \in [0, 1]$$

whenever $h : [0, 1] \times \Omega \to \mathcal{R}^n$ and $y : [0, 1] \to \mathcal{R}^n$ are continuous and $y(t) \notin h(t, \partial\Omega)$ for all $t \in [0, 1]$.

In principle, it is inessential how to introduce degree theory, since there is only one II-valued function d satisfying (d1), (d2), and (d3).

Therefore we refer to an excellent book by K. DEIMLING, "Nonlinear Functional Analysis", and start with the following theorem, which collects the properties of the Brouwer degree.

Definition 7.1 *There exists a unique II-valued mapping d, which associates every admissible triple (f, Ω, y), where $\Omega \subset \mathcal{R}^n, f : \overline{\Omega} \to \mathcal{R}^n$ is continuous and $y \in \mathcal{R}^n \setminus f(\partial\Omega)$ an integer $d(f, \Omega, y)$, with the following properties :*

The Topological Degree

1. If $d(f, \Omega, y) \neq 0$, then there exist $x \in \overline{\Omega}$ such that $f(x) = y$.

2. $d(id, \Omega, y) = 1$ if $y \in \Omega$, $\quad d(id, \Omega, y) = 0$, if $y \notin \Omega$.

$$[NORMALIZATION]$$

3. Let $h : [0, 1] \times \overline{\Omega} \to \mathcal{R}^n$ be continuous and $y \notin h(t, \partial\Omega)$ for $t \in [0, 1]$, then $d(h(t, .), \Omega, y)$ is independent from t.

$$[HOMOTOPY\ INVARIANCE]$$

4. If $g : \overline{\Omega} \to \mathcal{R}^n$ is continuous and $\|f - g\| < dist(y, f(\partial\Omega))$, then

$$d(f, \Omega, y) = d(g, \Omega, y).$$

5. If $z \in \mathcal{R}^n, \|y - z\| < dist(y, f(\partial\Omega))$, then

$$d(f, \Omega, y) = d(f, \Omega, z).$$

6. If $\cup_{i=1}^m \Omega_i \subset \Omega, \cup_{i=1}^m \overline{\Omega}_i \subset \overline{\Omega}, \Omega_i$ is open, disjoint, $y \notin \cup_{i=1}^m f(\partial\Omega_i)$, then

$$d(f, \Omega, y) = \sum_{i=1}^m d(f, \Omega_i, y).$$

$$[ADDITIVITY]$$

7. If $g : \overline{\Omega} \to \mathcal{R}^n$ is continuous and $f|_{\partial\Omega} = g|_{\partial\Omega}$ then

$$d(f, \Omega, y) = d(g, \Omega, y).$$

8. If A is a closed subset of $\overline{\Omega}, A \neq \overline{\Omega}$ and $y \notin f(A)$, then

$$d(f, \Omega, y) = d(f, \Omega \setminus A, y).$$

$$[EXCISION\ PROPERTY]$$

9. $d(f, \Omega, y) = d(f(.) - y, \Omega, 0)$.

80   Nonlinear Functional Analysis

10. *Let $m < n, \Omega \subset \mathcal{R}^n$ be open and bounded and $f : \overline{\Omega} \to \mathcal{R}^n$ continuous, $y \in \mathcal{R}^m \setminus (I - f)(\partial\Omega)$. Then*

$$d(I - f, \Omega, y) = d(I - f|_{\overline{\Omega} \cap \mathcal{R}^m}, \Omega \cap \mathcal{R}^m, y).$$

[REDUCTION]

7.2   Applications of the Brouwer Degree

Theorem 7.1 *(Brouwer)*

Let $D \subset \mathcal{R}^n$ be a nonempty compact convex set and let $f : D \to D$ be continuous. Then f has a fixed point. The same is true if D is only homeomorphic to a compact convex set.

Proof :

Suppose first that $D = \overline{B}_r(0)$. We may assume that $f(x) \neq x$ on ∂D. Let $h(t,x) = x - tf(x)$. This defines a continuous $h : [0,1] \times D \to \mathcal{R}^n$ such that $0 \notin h([0,1] \times \partial D)$, since by assumption $|h(t,x)| \geq |x| - t|f(x)| \geq (1-t)r > 0$ in $[0,1) \times \partial D$ and $f(x) \neq x$ for $|x| = r$. Therefore $d(id - f, D, 0) = d(id, B_r(0), 0) = 1$, and this proves existence of an $x \in B_r(0)$ such that $x - f(x) = 0$.

Next, let D be a general compact and convex set. By Theorem 5.4 we have a continuous extension $\tilde{f} : \mathcal{R}^n$ of f, and if we look at the defining formula in the proof of this result, we see that $\tilde{f}(\mathcal{R}^n) \subset \overline{conv f(D)} \subset D$ since

$$\left[\sum_{i=1}^m 2^{-i}\varphi_i(x)\right]^{-1} \sum_{i=1}^m w^{-i}\varphi_i(x)f(a^i)$$

is defined for $m = m(x)$ being sufficiently large, and belongs to conv $f(D)$. Now, we choose a ball $\overline{B}_r(0) \supset D$, and find a fixed

The Topological Degree

81

point x of $\tilde{f}$ in $\overline{B}_r(0)$, by the first step. But $\tilde{f}(x) \in D$ and therefore $x = \tilde{f}(x) = f(x)$.

Finally, assume that $D = h(D_0)$ with D_0 compact convex and h a homeomorphism. Then $h^{-1}fh : D_0 \to D_0$ has a fixed point x by the second step and therefore $f(h(x)) = h(x) \in D$.

$\square$

Let us illustrate this important theorem by some examples.

Example 7.1 (Perron - Frobenius)

Let $A = (a_{ij})$ be an $n \times n$ matrix such that $a_{ij} \geq 0$ for all i, j. Then there exist $\lambda \geq 0$ and $x \neq 0$ such that $x_i \geq 0$ for every i and $Ax = \lambda x$. In other words, A has a nonnegative eigenvector corresponding to a nonnegative eigenvalue.

To prove this result, let

$$D = \left\{ x \in \mathcal{R}^n : x_i \geq 0 \text{ for all } i \text{ and } \sum_{i=1}^{m} x_i = 1 \right\}.$$

If $Ax = 0$ for some $x \in D$, then there is no need to prove this result, with $\lambda = 0$. If $Ax \neq 0$ in D, then $\sum_{i=1}^{n}(Ax)_i \geq \alpha$ in D for some $\alpha > 0$. Therefore, $f : x \to Ax/\sum_{i=1}^{n}(Ax)_i$ is continuous in D, and $f(D) \subset D$ since $a_{ij} \geq 0$ for all i, j. By Theorem 7.2 we have a fixed point of f, i.e. an $x_0 \in D$ such that $Ax_0 = \lambda x_0$ with $\lambda = \sum_{i=1}^{n}(Ax_0)_i$.

Example 7.2

Lets consider the system of ordinary differential equations $u' = f(t, u)$, where $u' = \frac{du}{dt}$ and $f : \mathcal{R} \times \mathcal{R}^n \to \mathcal{R}^n$ is ω - periodic solutions. Suppose, for simplicity, that f is continuous and that there is a ball $B_r(0)$ such that the initial value problems

$$u' = f(t, u), \quad u(0) = x \in \overline{B}_r(0) \tag{7.1}$$

have a unique solution $u(t; x)$ on $[0, \infty)$.

Now, let $p_t x = u(t; x)$ and suppose also that f satisfies the boundary condition $(f(t,x), x) = \sum_{i=1}^n f_i(t,x)x_i < 0$ for $t \in [0, \omega]$ and $|x| = r$. Then, we have $P_t : \overline{B}_r(0) \to \overline{B}_r(0)$ for every $t \in \mathcal{R}^+$, since

$$\frac{d}{dt}|u(t)|^2 = 2(u'(t), u(t)) = 2(f(t, u(t)), u(t)) < 0$$

if the solution u of equation (7.1) takes a value in $\partial B_r(0)$ at time t. Furthermore, P_t is continuous, as follows easily from our assumption that equation (7.1) has only one solution. Thus we find P_ω has a fixed point $x_\omega \in \overline{B}_r(0)$, i.e. $u' = f(t, u)$ has a solution such that $u(0; x_\omega) = x_\omega = u(\omega, x_\omega)$. Now, we may easily check that $v : [0, \infty) \to \mathcal{R}^n$, defined by $v(t) = u(t - k\omega, x_\omega)$ on $[k\omega, (k+1)\omega]$, is an ω - periodic solution of equation (7.1). The map P_ω is usually called the Poincare operator of $u' = f(t, u)$, and it is now evident that $u(.; x)$ is an ω - periodic solution iff x is a fixed point of P_ω.

Example 7.3

It is impossible to retract the whole unit ball continuously onto its boundary such that the boundary remains pointwise fixed, i.e. there is no continuous $f : \overline{B}_1(0) \to \partial B_1(0)$ such that $f(x) = x$ for all $x \in \partial B_1(0)$.

Otherwise $g = -f$ would have a fixed point x_0, by Theorem 7.2, but this implies $|x_0| = 1$ and therefore $x_0 = -f(x_0) = -x_0$, which is nonsense. This result is in fact equivalent to Brouwer's theorem for the ball. To see this, suppose $f : \overline{B}_1(0) \to \overline{B}_1(0)$ is continuous and has no fixed point. Let $g(x)$ be the point where the line segment from $f(x)$ to x hits $\partial B_1(0)$, i.e. $g(x) = f(x) + t(x)(x - f(x))$, where $t(x)$ is the positive root of

$$t^2|x - f(x)|^2 + 2t(f(x), x - f(x)) + |f(x)|^2 = 1.$$

The Topological Degree 83

Since $t(x)$ is continuous, g would be such a retraction which does not exist by assumption.

Surjective Maps

In this section we shall show that a certain growth condition of $f \in C(\mathcal{R}^n)$ implies $f(\mathcal{R}^n) = \mathcal{R}^n$. Let us first consider that $f_0(x) = Ax$ with a positive definite matrix A. Since $\det A \neq 0$, f_0 is surjective. We also have $(f_0(x), x) \geq c|x|^2$ for some $c > 0$ and every $x \in \mathcal{R}^n$, and therefore $(f_0(x), x)/|x| \to \infty$ as $|x| \to \infty$. This condition is sufficient for surjectivity in the nonlinear case too, since we can prove the following theorem.

Theorem 7.2 *Let $f \in C(\mathcal{R}^n)$ be such that $(f_0(x), x)/|x| \to \infty$ as $|x| \to \infty$. Then $f(\mathcal{R}^n) = \mathcal{R}^n$.*

Proof :

Given $y \in \mathcal{R}^n$, let $h(t, x) = tx + (1 - t)f(x) - y$. At $|x| = r$ we have

$$(h(t, x), x) \geq r[tr + (1 - t)(f(x), x)/|x| - |y|] > 0$$

for $t \in [0, 1]$ and $r > |y|$ being sufficiently large. Therefore, $d(f, B_r(0), y) = 1$ for such an r, i.e. $f(x) = y$ has a solution.

$\square$

Hedgehog Theorem

Up to now we have applied the homotopy invariance of the degree as it stands. However, it is also useful to use the converse namely: if two maps f and g have different degree then a certain h that connects f and g cannot be a homotopy. Along these lines we shall prove the following theorem.

Theorem 7.3 *Let $\Omega \subset \mathcal{R}^n$ be open bounded with $0 \in \Omega$ and let $f : \partial\Omega \to \mathcal{R}^n \setminus \{0\}$ be continuous. Suppose also that the space*

84   *Nonlinear Functional Analysis*

dimension n is odd. Then there exist $x \in \partial\Omega$ and $\lambda \neq 0$ such that $f(x) = \lambda x$.

Proof :

Without loss of generality we may assume $f \in C(\overline{\Omega})$, by Proposition 5.1. Since n is odd, we have $d(-id, \Omega, 0) = -1$. If $d(f, \Omega, 0) \neq -1$, then $h(t, x) = (1 - t)f(x) - tx$ must have a zero $(t_0, x_0) \in (0, 1) \times \partial\Omega$. Therefore, $f(x_0) = t_0(1 - t_0)^{-1}x_0$. If, however, $d(f, \Omega, 0) = -1$ then we apply the same argument to $h(t, x) = (1 - t)f(x) + tx$.

$\square$

Since the dimension is odd in this theorem, it does not apply in C^n. In fact, the rotation by $\frac{\pi}{2}$ of the unit sphere in $C = R^2$, i.e. $f(x_1, x_2) = (-x_2, x_1)$, is a simple counter example. In case $\Omega = B_1(0)$ the theorem tells us that there is at least one normal such that f changes at most its orientation. In other words: there is no continuous nonvanishing tangent vector field on $S = \partial B_1(0)$, i.e. an $f : S \to R^n$ such that $f(x) \neq 0$ and $(f(x), x) = 0$ on S. In particular, if $n = 3$ this means, that 'a hedgehog cannot be combed without leaving tufts or whorls'. However, $f(x) = (x_2, -x_1, ..., x_{2m}, -x_{2m-1})$ is a nonvanishing tangent vector field on $S \subset R^{2m}$.

The proof of existence and uniqueness of the Brouwer degree and its construction is based on the fact, that $f(S_f(\Omega))$ is of measure zero, where $S_f(\Omega)$ is the set of critical points of f, i.e. $S_f(\Omega) = \{x \in \Omega, J_f(x) = det f'(x) = 0\}$ (Sard's lemma), and approximations of continuous functions by differentiable functions.

Proposition 7.1 *Let $\Omega \subset R^n$ be open and $f \in C^1(\Omega)$. Then $\mu_n(f(S_f)) = 0$, where μ_n denotes the n - dimensional Lebesgue measure.*

The Topological Degree

Proof :

We need to know here about Lebesgue measure μ_n is that $\mu_n(J) = \prod_{i=1}^{n}(b_i - a_i)$ for the interval $J = [a,b] \subset \mathcal{R}^n$ and that $M \subset \mathcal{R}^n$ has measure zero (i.e. $\mu_n(M) = 0$) iff to every $\epsilon > 0$ there exist at most countably many intervals J_i such that $M \subset \cup_i J_i$ and $\sum_i \mu_n(J_i) \leq \epsilon$.

Then it is easy to see that at most countable union of sets of measure zero also has measure zero.

Since an open set Ω in $\mathcal{R}^n$ may be written as a countable union of cubes, say $\Omega = \cup_i Q_i$, it is therefore sufficient to show $\mu_n(f(S_f(Q))) = 0$ for a cube $Q \subset \Omega$, since $f(S_f(\Omega)) = \cup_i f(S_f(Q_i))$. Let ρ be the lateral length of Q. By the uniform continuity of f' on Q, given $\epsilon > 0$, we then find $m \in \mathcal{N}$ such that $|f'(x) - f'(\overline{x})| \leq \epsilon$ for all $x, \overline{x} \in Q$ with $|x - \overline{x}| \leq \delta = \sqrt{n}.\frac{\rho}{m}$, and therefore

$$|f(x) - f(\overline{x}) - f'(\overline{x})(x - \overline{x})| \leq \int_0^1 |f'(\overline{x} - t(x - \overline{x})) - f'(\overline{x})||x - \overline{x}|dt \leq \epsilon|x - \overline{x}|$$

for any such $x, \overline{x}$. So let us decompose Q into r cubes Q^k of diameter δ. Since $\delta/\sqrt{n}$ is the lateral length of Q^k, we have $r = m^n$ and

$$f(x) = f(\overline{x}) + f'(\overline{x})(x - \overline{x}) + R(x, \overline{x})$$

with

$$|R(x, \overline{x})| \leq \epsilon\delta \text{ for } x, \overline{x} \in Q^k.$$

Now, assuming that $Q^k \cap S_f \neq \phi$, choose $\overline{x} \in Q^k \cap S_f$; let $A = f'(\overline{x})$ and $g(y) = f(\overline{x} + Y) - F(\overline{x})$ for $y \in \overline{Q}^k = Q^k - \overline{x}$. Then we have

$$g(y) = Ay + \overline{R}(y)$$

86 Nonlinear Functional Analysis

with

$$|\tilde{R}(y)| = |R(\overline{x} + y, \overline{x})| \le \epsilon\delta \text{ on } \tilde{Q}^k.$$

Since $\det A = 0$, we know that $A(\tilde{Q}^k)$ is contained in a $(n-1)$ - dimensional subspace of $\mathcal{R}^n$. Hence, there exists $b^1 \in \mathcal{R}^n$ with $|b^1| = 1$ and $(x, b^1) = \sum_{i=1}^n x_i b_{\frac{1}{2}} = 0$ for all $x \in A(\tilde{Q}^k)$. Extending b^1 to an orthonormal base $\{b^1, ..., b^n\}$ of $\mathcal{R}^n$, we have

$$g(y) = \sum_{i=1}^n (g(y), b^i) b^i$$

with

$$|(g(y), b^1)| \le |\tilde{R}(y)||b^1| \le \epsilon\delta$$

and

$$|g(y), b^i)| \le |A||y| + |\tilde{R}(y)| \le |A|\delta + \epsilon\delta \text{ for } i = 2, ..., n,$$

where $|A| = |(a_{ij})| = \left(\sum_{i,j=1}^n a_{ij}^2\right)^{\frac{1}{2}}$. Thus, $f(Q^k) = f(\overline{x}) + g(\tilde{Q}^k)$ is contained in an interval J_k around $f(\overline{x})$ satisfying

$$\mu_n(J_k) = [2(|A|\delta + \epsilon\delta)]^{n-1} \cdot 2\epsilon\delta = 2^n(|A| + \epsilon)^{n-1}\epsilon\delta^n.$$

Since f' is bounded on the large cube Q, we have $|f'(x)| \le c$ for some c, in particular $|A| \le c$. Therefore, $f(S_f(Q)) \subset \cup_{k=1}^r J_k$ with

$$\sum_{k=1}^r \mu_n(J_k) \le r.2^n(c + \epsilon)^{n-1}\epsilon\delta^n = 2^n(c + \epsilon)^{n-1}(\sqrt{n}\rho)^n\epsilon,$$

i.e. $f(S_f(Q))$ has measure zero, since $\epsilon > 0$ is arbitrary.

$\square$

Having this result we give the following definition.

Definition 7.2 (a) Let $\Omega \subset \mathcal{R}^n$ be open and bounded, and let $f_0 : \overline{\Omega} \to \mathcal{R}^n$ be continuous and twice continuously differentiable

The Topological Degree 87

in Ω. Let $y_0 \in \mathcal{R}^n \setminus f_0(\partial\Omega \cup S_{f_0}(\Omega))$. Then

$$d(f_0, \Omega, y_0) = \sum_{x \in f_0^{-1}\{y\}} sgn J_{f_0}(x).$$

(b) Let $f : \overline{\Omega} \to \mathcal{R}^n$ be continuous and $y \notin f(\partial\Omega)$. Choose a twice continuously differentiable function $f_0 : \overline{\Omega} \to \mathcal{R}^n$, and $y_0 \notin f_0(\partial\Omega \cup S_{f_0}(\Omega))$, such that $\|f - f_0\| < r$, $\|y - y_0\| < r$, where

$$r = dist(y, f(\partial\Omega)).$$

Then by (a) we set

$$d(f, \Omega, y) = d(f_0, \Omega, y_0).$$

7.3 The Leray-Schauder Degree

Let X be a real Banach space, $\Omega \subset X$ an open and bounded subset of X, $F : \overline{\Omega} \to X$ compact and $y \notin (I - F)(\partial\Omega)$. On these admissible triplets we want to define a $\amalg$- valued function D that satisfies the three basic conditions corresponding to (d1), (d2) and (d3) of the Brouwer degree, namely,

(D1) $D(I, \Omega, y) = 1$ for $y \in \Omega$

(D2) $D(I - F, \Omega, y) = D(I - F, \Omega_1, y) + D(I - F, \Omega_2, y)$
whenever Ω_1 and Ω_2 are disjoint open subsets of Ω such that $y \notin (I - F)(\overline{\Omega} \setminus (\Omega_1 \cup \Omega_2))$

(D3) $D(I - H(t, .), \Omega, y(t))$ is independent of $t \in [0, 1]$, whenever $H : [0, 1] \times \overline{\Omega} \to X$ is compact, $y : [0, 1] \to$ is continuous and $y(t) \notin (I - H(t, .))(\partial\Omega)$ on $[0, 1]$.

In the first step we will extend the Brouwer degree to finite dimensional Banach spaces. Let X be a n - dimensional real

Banach space, $\Omega \subset X$ open and bounded, and $F : \overline{\Omega} \to X$ continuous. Let $\{x_1, ..., x_n\}$ be a basis in $X, \{e_1, ..., e_n\}$ the unit vector basis of $\mathcal{R}^n$, and $h : X \to \mathcal{R}^n$ the linear homeomorphism, defined by

$$h(x_j) = e_j, \quad j = 1, 2, ..., n.$$

Then $h(\Omega)$ is open and bounded in $\mathcal{R}^n$, $f = hFh^{-1} : \overline{h(\Omega)} \to \mathcal{R}^n$ is continuous and $h(y) \notin f(\partial h(\Omega))$. Then we define

$$d_x(F, \Omega, y) = d(hFh^{-1}, h(\Omega), h(y)).$$

It is easy to check by the properties of the Brouwer degree that this definition is independent of the choice of the basis of X.

In the second step we will assume, that X is an arbitrary real Banach space, Ω an open and bounded subset of X and $F : \overline{\Omega} \to X$ a compact map. By $\mathcal{F}(\overline{\Omega})$ we denote the compact mappings $F : \overline{\Omega} \to X$ such that $F(\overline{\Omega})$ is contained in a finite dimensional subspace of X. Let $G = I - F$ and $y \notin X \setminus G(\partial\Omega)$. By Lemma 6.5 $G(\partial\Omega)$ is closed, and $r = \text{dist}(y, G(\partial\Omega)) > 0$.

Now let $F_1 \in \mathcal{F}(\overline{\Omega})$ be an approximation of F, such that $\sup_{x \in \Omega} \|F(x) - F_1(x)\| < r$. Then we have

$$y \notin (I - F_1)(\partial\Omega)$$

since

$$\begin{aligned}
\text{dist}(y, (I - F_1)(\partial\Omega)) &= \inf_{x \in \partial\Omega} \|y - (x - F_1(x))\| \\
&\geq \inf_{x \in \partial\Omega} (\|y - (x - F(x))\| - \|F_1(x) - F(x)\|) > 0.
\end{aligned}$$

Now we choose a finite dimensional subspace X_1 of X, such that $y \in X_1$, $\Omega \cap X_1 \neq \phi$ and $F_1(\overline{\Omega}) \subset X_1$. Then $\Omega_1 = \Omega \cap X_1$ is open and bounded in X_1, and by the first step we have the existence of

$$d_x(I - F_1)|_{\overline{\Omega}_1}, \Omega_1, y).$$

The Topological Degree

This number is independent of the choice of F_1 and X_1. Let $F_2 \in \mathcal{F}(\overline{\Omega})$, such that $\|F - F_2\| < R$ and $X_2 \subset X$, such that $\dim X_2 < \infty$ and $y \in X_2, F_2(\overline{\Omega}) \subset X_2, \Omega_2 = \Omega \cap X_2 \neq \phi$. Now let $X_0 = span\{X_1, X_2\}$ be the finite dimensional subspace of X generated by X_1 and $X_2, \Omega_0 = X_0 \cap \Omega$, then by the reduction property (1) of the Brouwer degree for $j = 1, 2$

$$d_x((I - F_j)|_{\overline{\Omega}_j}, \Omega_j, y) = d_x((I - F_j)|_{\overline{\Omega}_0}, \Omega_0, y).$$

The homotopy $H : [0, 1] \times \overline{\Omega}_0 \to X_0$ with

$$H(t, x) = t(x - F_1(x)) + (1 - t)(x - F_2(x))$$

connects $I - F_1$ with $I - F_2$ and it is

$$\|x - F(x) - H(t, x)\| = \|F(x) - tF_1(x) - (1 - t)F_2(x)\|$$
$$\leq t\|F(x) - F_1(x)\| + (1 - t)\|F(x) - F_2(x)\| < r,$$

hence $y \notin H([0, 1] \times \partial\Omega_0)$. Therefore by the homotopy invariance principle

$$d_x((I - F_1)|_{\overline{\Omega}_0}, \Omega_0, y) = d_x((I - F_2)|_{\overline{\Omega}_0}, \Omega_0, y).$$

Thus it is legitimate to give the following definition.

Definition 7.3 *(Leray - Schauder Degree)*

Let X be a real Banach space, $\Omega \subset X$ a bounded and open subset of $X, F : \overline{\Omega} \to X$ a compact map, $y \in X$ such that $y \notin (I - F)(\partial\Omega)$. Let

$$D(I - F, \Omega, y) = d_x((I - F_1)|_{\overline{\Omega}_1}, \Omega_1, y),$$

where $F_1 \in \mathcal{F}(\overline{\Omega})$ is chosen, such that with a finite dimensional subspace X_1 of X

$$\|F - F_1\| < dist(y, (I - F)(\partial\Omega))$$
$$F_1(\overline{\Omega}) \subset X_1, y \in X_1, \Omega_1 = \Omega \cap X_1 \neq \phi,$$

this number $D(I - F, \Omega, y)$ is called the Leray-Schauder degree.

90   Nonlinear Functional Analysis

Now it is rather clear that we obtain nearly all properties of the Brouwer degree for the Leray - Schauder degree, too.

Theorem 7.4 *Besides (D1), (D2), (D3), the Leray - Schauder degree has the following properties:*

(D4) $D(I - F, \Omega, y) \neq 0$ *implies* $(I - F)^{-1}\{y\} \neq \phi$.

(D5) $D(I - F, \Omega, y) = D(I - G, \Omega, y)$ *for* $G : \overline{\Omega} \to X$
        compact $\|F - G\| < r = dist(y, (I - F)(\partial\Omega))$.

(D6) $D(I - F, \Omega, y) = D(I - F, \Omega, z)$ *for* $z \in X$,
        $\|y - z\| < r = dist(y, (I - F)(\partial\Omega))$.

(D7) $D(I - F, \Omega, y) = D(I - G, \Omega, y)$, *whenever*
        $G|_{\partial\Omega} = F|_{\partial\Omega}$.

(D8) $D(I - F, \Omega, y) = D(I - F, \Omega_*, y)$ *for every open*
        subset Ω_* *of* Ω *such that* $y \notin (I - F)(\overline{\Omega} \setminus \Omega_*)$.

(D9)   *Let* Y *be a closed subspace of* $X, \Omega \subset X$ *be open and*
        bounded, $\Omega \cap Y \neq \phi$. *Let* $F : \overline{\Omega} \to Y$ *compact,*
        $y \notin Y \setminus (I - F)(\partial\Omega)$. *Then*
        $D(I - F, \Omega, y) = D((I - F)|_{\overline{\Omega} \cap Y}, \Omega \cap Y, y)$.

Proof :

(D4) Let $F_n \in \mathcal{F}(\overline{\Omega})$ such that $\|F - F_n\| < \frac{1}{n}r$. Choose X_n, Ω_n according to the Definition 7.7. Then

$$d_x((I - F_n)|_{\overline{\Omega}_n}, \Omega_n, y) \neq 0$$

i.e. $\exists x_n \in \Omega_n$ such that $x_n - F_n(x_n) = y$. Thus

$$\|x_n - F(x_n) - y\| \leq \|x_n - F_n(x_n) - y\| + \|F(x_n) - F_n(x_n)\| \leq \frac{1}{n}r \to 0.$$

Since $(I - F)(\overline{\Omega})$ is closed, $y \in (I - F)(\overline{\Omega})$, hence $x - F(x) = y$ for an $x \in \overline{\Omega}, x \notin \partial\Omega$, i.e. $x \in \Omega$.

The Topological Degree 91

(D5) Choose $G_1 \in \mathcal{F}(\overline{\Omega})$ such that

$$\|G - G_1\| < \min\{\operatorname{dist}(y, (I - G)(\partial\Omega)),$$
$$\operatorname{dist}(y, (I - F)(\partial\Omega)) - \|F - G\|\}.$$

Then

$$D(I - G, \Omega, y) = d_x((I - G_1)|_{\overline{\Omega}_1}, \Omega_1, y).$$

From

$$\|G_1 - F\| \le \|G - F\| + \|G - G_1\| < \operatorname{dist}(y, (I - F)(\partial\Omega))$$

follows $D(I - F, \Omega, y) = d_x(I - G_1)|_{\overline{\Omega}_1}, \Omega_1, y)$.

(D6) is similar to (D5).

(D7) Let $H(t, x) = tF(x) + (1 - t)G(x)$. Then
$y \notin H([0, 1] \times \partial\Omega)$ if $F|_{\partial\Omega} = G|_{\partial\Omega}$.
Thus $D(I - F, \Omega, y) = D(I - G, \Omega, y)$.

(D8) $(I - F)(\overline{\Omega} \setminus \Omega_*)$ is closed.
Let $r = \operatorname{dist}(y, (I - F)(\overline{\Omega} \setminus \Omega_*)) > 0$ and $F_1 \in \mathcal{F}(\overline{\Omega})$ such that
$\|F - F_1\| < \frac{r}{2}$. Then

$$\operatorname{dist}(y, (I - F_1)(\overline{\Omega} \setminus \Omega_*)) \ge r - \|F - F_1\| > \frac{r}{2}$$

and for properly chosen $X_1 \subset X, \Omega_1 = \Omega \cap X_1$

$$
\begin{aligned}
D(F, \Omega, y) &= d_x((I - F_1)|_{\overline{\Omega}_1}, \overline{\Omega}_1, y) \\
&= d_x((I - F_1)|_{\overline{\Omega}_1}, \Omega_* \cap X_1, y) \\
&= D(F, \Omega_*, y).
\end{aligned}
$$

(D9) Since $F(\overline{\Omega}) \subset Y$ for $r = \operatorname{dist}(y, (I - F)(\partial\Omega))$ there
exists $F_1 \in \mathcal{F}(\overline{\Omega}), \|F - F_1\| < r$ and $F_1(\overline{\Omega}) \subset Y$. Let $X_1 \subset X$,

$\dim X_1 < \infty, y \in X_1, \Omega \cap Y \cap X_1 \neq \phi$. Then

$$
\begin{aligned}
D(I - F, \Omega, y) &= d_x(I - F_1|_{\overline{\Omega} \cap X_1}, \Omega \cap X_1, y) \\
&= d_x(I - F_1|_{\overline{\Omega} \cap Y \cap X_1}, \Omega \cap Y \cap X_1, y) \\
&= D(I - F|_{\overline{\Omega} \cap Y}, \Omega \cap Y, y)
\end{aligned}
$$

by reduction property (1) of Theorem 7.1.

$\square$

7.4 Borsuk's Antipodal Theorem

Whenever we want to show by means of degree theory that $f(x) = y$ has a solution in Ω, we have to verify $d(f, \Omega, y) \neq 0$. For symmetric domains Ω (with respect to the origin) and for odd maps Borsuk's theorem states that $d(f, \Omega, 0)$ is odd, hence different from zero. We say that $\Omega \subset X$ is **symmetric** with respect to the origin, if $\Omega = -\Omega$, and $f : \Omega \to X$ is said to be **odd** if $f(-x) = -f(x)$ for $x \in \Omega$. We start with $X = \mathcal{R}^n$.

Theorem 7.5 Let $\Omega \subset \mathcal{R}^n$ be open, bounded, symmetric with $0 \in \Omega$. Let $f : \overline{\Omega} \to \mathcal{R}^n$ be odd and $0 \notin f(\partial\Omega)$. Then $d(f, \Omega, 0)$ is odd.

Proof :

1. We may assume that f is continuously differentiable in Ω and $j_f(0) \neq 0$. To see this, approximate f by a differentiable g_1 and consider g_2 with $g_2(x) = \frac{1}{2}(g_1(x) - g_1(-x))$, choose δ which is not an eigenvalue of $g_2'(0)$. Then $\tilde{f}(x) = g_2(x) - \delta - x$ is continuously differentiable, odd, $I_{\tilde{f}}(x) \neq 0$, and

$$
\begin{aligned}
\|f - \tilde{f}\| &= \sup_{\Omega} |\frac{1}{2}(f(x) - g_1(x)) - \frac{1}{2}(f(-x) - g_1(x)) - \delta x| \\
&\leq \|f - g_1\| + \delta \text{ diam } \Omega
\end{aligned}
$$

The Topological Degree

can be chosen arbitrarily small, hence

$$d(f, \Omega, 0) = d(\tilde{f}, \Omega, 0).$$

2. Let f be continuously differentiable and $J_f(0) \neq 0$. Now to prove the theorem it suffices to show that there is an odd $g : \overline{\Omega} \to \mathcal{R}^n$, differentiable, close to f such that $0 \notin g(S_g(\Omega))$, since then

$$d(f, \Omega, 0) = d(g, \Omega, 0) = sgn J_g(0) + \sum_{\substack{x \neq 0 \\ x \in g^{-1}\{0\}}} sgn J_g(x)$$

where the sum is even, since $g(x) = 0$ iff $g(-x) = 0$ and $J_g(x)$ is even, since $g(x) = -g(-x)$ implies $g'(x) = g'(-x)$.

3. We begin with the following observation. Let

$$u : \Omega \to \mathcal{R}^n, \; v : \Omega \to \mathcal{R}$$

be continuously differentiable, $v(x) \neq 0$ if x belongs to an open subset A of Ω. For $y \in \mathcal{R}^n$ define

$$h_y(x) = u(x) - v(x).y,$$

then y is a regular value of $\frac{u}{v}|_A$ iff 0 is a regular value of $h_y|_A$, since for $x \in A, \frac{u}{v}(x) = y$, thus $h_y(x) = 0$ implies by the quotient rule $(\frac{u}{v})'(x) = \frac{1}{v(x)}h'_y(x)$, because

$$(\frac{u}{v})'(x) = \frac{u'(x)}{v(x)} \frac{v'(x)u(x)}{v(x)^2} = \frac{1}{v(x)}.[u'(x) - v'(x).\frac{u}{v}(x)]$$

$$= \frac{1}{v(x)}(u'(x) - v'(x)y),$$

thus (by Sard's lemma)

(*) 0 is a regular value of $h_y|_A$ for almost all $y \in \mathcal{R}^n$.

94   *Nonlinear Functional Analysis*

Now we define odd functions $h_1, ..., h_n \in \overline{C}^1(\Omega)$ with $\|f - h_k\| < \epsilon$ if $x \in \overline{\Omega}$ and 0 is regular value of $h_k|_{\Omega_k}$, where

$$H_k = \{x = (\xi_1, ..., \xi_n) \in \mathcal{R}^n, \xi_k = 0\}$$
$$\Omega_k = \Omega \setminus (H_1 \cap ... \cap H_k).$$

Let $h_1(x) = f(x)\xi_1^3 y_1$, where $y_1 \in \mathcal{R}^n$, such that 0 is a regular value of $h_1|_{\Omega_1}$ and $|y_1|$ sufficiently small.

If h_k is defined, we define h_{k+1} by $h_{k+1}(x) = h_k(x) - \xi_{k+1}^3 \cdot y_{k+1}$ with a sufficiently small y_{k+1}, such that 0 is a regular value of $h_{k+1}|_{\Omega \setminus H_{k+1}}$. If $x \in \Omega \cap H_{k+1}$ we have $h_{k+1}(x) = h_k(x)$ and $h'_{k+1}(x) = h'_k(x)$, for $x \in \Omega_k \cap H_{k+1}$, by induction we get : $h_{k+1}(x) = 0$, then $h'_{k+1}(x)$ is regular.

Since

$$(\Omega_k \cap H_{k+1}) \cup (\Omega \setminus H_{k+1})$$
$$= (\Omega \setminus h_1 \cap ... \cap H_k) \cap H_{k+1} \cup (\Omega \setminus H_{k+1})$$
$$= \Omega \setminus H_1 \cap ... \cap H_{k+1} = \Omega_{k+1}$$

we have that 0 is a regular value of $h_{k+1}|_{\Omega_{k+1}}$. Let $g = h_n$, then 0 is a regular value of $h_n|_{\Omega \setminus \{0\}}$, but by definition of $g = h_n$ we obtain $g'(0) = f'(0), 0$ is a regular point of g.

$\square$

Corollary 7.1 *Let $\Omega \subset \mathcal{R}^n$ be open, bounded, symmetric and $0 \in \Omega$. Let $f : \overline{\Omega} \to \mathcal{R}^n$ be continuous such that for all $\lambda \geq 1$ and $x \in \partial\Omega$*

$$f(x) \neq 0, \quad f(-x) - \lambda f(x) \neq 0.$$

Then $d(f, \Omega, 0)$ is odd.

The Topological Degree

95

Proof :

The mapping $h(t, x) = f(x) - t(-x)$ defines a homotopy in $\mathcal{R}^n \setminus \{0\}$ between f and the odd function $g(x) = f(x) - f(-x)$.

$\square$

Corollary 7.2 *(Borsuk - Ulam)*

Let $\Omega \subset \mathcal{R}^n$ be as before, $f : \partial\Omega \to \mathcal{R}^m$ be continuous and $m < n$. Then there exists an $x \in \partial\Omega$ such that

$$f(x) = f(-x).$$

Proof :

If not, $g(x) = f(x) - f(-x) \neq 0$ for all $x \in \partial\Omega$. Again by g we denote a continuous extension to $\overline{\Omega}$ of these boundary values. Then $d(g, \Omega, y) = d(g, \Omega, 0)$ for all y in a properly chosen neighbourhood U of 0 (by (5) of definition (7.1)).

Thus $U \subset g(\overline{\Omega}) \subset \mathcal{R}^m$, which is not possible.

$\square$

Theorem 7.6 Let $\Omega \subset \mathcal{R}^n$ be open, bounded and symmetric with respect to $0 \in \Omega$ and $\{A_1, ..., A_p\}$ be a covering of $\partial\Omega$ by closed sets $A_j \subset \partial\Omega$ such that $A_j \cap (-A_j) = 0$ for $j = 1, 2, ..., p$. Then $p \geq n + 1$.

Proof :

Suppose that $p \leq n$, let $f_j(x) = 1$ on A_j and $f_j(x) = -1$ on $-A_j$ for $j = 1, 2, ..., p-1$ and $f_j(x) = 1$ on Ω for $j = p, p+1, ..., n$.

Extend $f_1, ..., f_{p-1}$ continuously to $\overline{\Omega}$. Let $x \in A_p$, then $-x \notin A_p$, and therefore $-x \in A_j$ for some $j \leq p - 1$, i.e. $x \in -A_j$. Hence

$$\partial\Omega \subset \cup_{j=1}^{p-1}(A_j \cup (-A_j)).$$

Let $x \in A_j$, then $f_j(x) = 1, f_j(-x) = -1$ and $x \in -A_j$ implies $f_j(x) = -1, f_j(-x) = 1$. Let $f = (f_j)$, then $f(-x)$ does not point in the same direction as $f(x)$, thus $f(-x) \neq \lambda f(x)$ on $\partial\Omega$ for $\lambda > 0$, by Corollary 7.1 $d(f, \Omega, 0)$ is odd, i.e. there exists an $x \in \Omega$ such that $f(x) = 0$, which contradicts $f_n(x) = 1$.

$\square$

Now we are able to compute the measure of noncompactness of the unit ball.

Corollary 7.3 *Let* $B = B(0,1)$ *be the unit ball in an infinite dimensional Banach space* X. *Then the measure of noncompactness* $\chi(B) = 2$.

Proof :

Let $S = \partial B$ be the boundary of B and

$$S = \cup_{j=1}^n M_j$$

be a finite covering of S by closed sets $M_j \subset S$ such that $\text{diam}(M_j) < 2$. Let X_n be an n - dimensional subspace of X, then $S \cap X_n = \cup_{j=1}^n (M_j \cap X_n)$ is the boundary of the unit ball in X_n, and therefore one of the $M_j \cap X_n$ contains a pair of antipodal points x and $-x$ by theorem 7.6. Hence

$$\text{diam } M_j \geq ||x - (-x)|| = 2||x|| = 2.$$

This contradiction along with the fact $\chi(B) \leq 2$ shows $\chi(B) = 2$.

$\square$

If X is infinite dimensional, the Borsuk theorem follows immediately from the finite dimensional case:

The Topological Degree

Theorem 7.7 *Let $\Omega \subset X$ be open, bounded and symmetric with respect to $0 \in \Omega, F : \overline{\Omega} \to X$ be compact, $G = I - F$ and $0 \notin G(\partial\Omega)$. If for all $\lambda \geq 1$ and for all $x \in \partial\Omega$, $G(-x) - \lambda G(x) \neq 0$, then $D(I - F, \Omega, 0)$ is odd. In particular, this is true if $F|_{\partial\Omega}$ is odd.*

Proof :

Let $H(t, x) = \frac{t}{1+t}.F(x) - \frac{1}{1+t}.F(-x)$, then H is compact and $x \neq H(t, x)$ for all $x \in \partial\Omega, t \in [0, 1]$ since

$$
\begin{aligned}
x - H(t, x) &= \frac{t}{1+t}.(\frac{1}{t}(1+t)x - \frac{1}{t}F(x) + F(-x)) \\
&= \frac{t}{1+t}.(\frac{1}{t}(x - F(x)) - (-x - F(-x))) \\
&= \frac{t}{1+t}.(\frac{1}{t}G(x) - G(-x)) \neq 0.
\end{aligned}
$$

Let $F_0(x) = \frac{1}{2}(F(x) - F(-x))$, then F_0 is odd and thus $D(I - F, \Omega, 0) = D(I - F_0, \Omega, 0)$. Choose $F_1 \in \mathcal{F}(\overline{\Omega})$ with $\|F_0 - F_1\| < \text{dist}(0, (I - F_0)(\partial\Omega))$ and $F_2(x) = \frac{1}{2}(F_1(x) - F_1(-x))$, then F_2 is odd and $\|F_0 - F_2\| \leq \|F_0 - F_1\|$, therefore

$$
D(I - F_0, \Omega, 0) = d((I - F_2)|_{\overline{\Omega}}, \Omega_2, 0) \text{ is odd.}
$$

$\square$

We will conclude this section with the following corollaries.

Corollary 7.4 *Let $\Omega \subset X$ be open, bounded and symmetric with respect to $0 \in \Omega$, and $F : \overline{\Omega} \to X$ compact and odd. Let $Y \subset X$ be a proper linear subspace and $(I - F)(\overline{\Omega}) \subset Y$. Then there exists a fixed point $x_0 \in \partial\Omega$ of F.*

Nonlinear Functional Analysis

Proof :

If $0 \notin (I - F)(\partial \Omega)$, then $D(I - F, \Omega, 0)$ is odd, i.e. for all $y \in \Delta$, then connected component of $X \setminus (I - F)(\partial \Omega)$, which contains 0, there exists an $x \in \Omega$, such that $x - F(x) = y$, i.e. $\Delta \subset (I - F)(\Omega) \subset Y \stackrel{C}{\neq} X$. This contradicts the openness of Δ.

$\square$

Corollary 7.5 *Let $\Omega \subset X$ be open, bounded and symmetric with respect to $0 \in \Omega$, and $F : \overline{\Omega} \to X$ compact. Let $(I-F)(\overline{\Omega}) \subset Y \stackrel{C}{\neq} X$. Then there is an $x \in \partial \Omega$ such that*

$$x - F(x) = -x - F(-x).$$

Proof :

Let $G(x) = \frac{1}{2}(F(x) - F(-x))$. By Corollary 7.4 there is an $x \in \partial \Omega$ with $G(x) = x$, thus $\frac{1}{2}(x - F(x)) + \frac{1}{2}(x + F(-x)) = 0$ or $x - F(x) = -x - F(-x)$.

$\square$

Corollary 7.6 *Let $F : X \to X$ be compact, and $L : X \to X$ compact and linear. If there exists a $\lambda \in \mathcal{R}$, such that for all $x \in X, ||x|| = r$ we have*

$$||F(x) - \lambda L x|| < ||x - \lambda L x||,$$

then $D(I - F, B(0, r), 0)$ is odd.

Proof :

Let $H : [0, 1] \times B(0, r) \to X$ be defined by

$$H(t, x) = x - (1 - t)\lambda L x - t F(x),$$

The Topological Degree

then

$$||H(t,x)|| \geq ||x - \lambda Lx|| - t||F(x) - \lambda Lx|| > 0,$$

hence $0 \notin H([0,1] \times \partial\Omega)$. By homotopy invariance of the degree

$$D(I - F, B(0,r), 0) = D(I - \lambda L, b(0,r), 0)$$

is odd, since linear opeators are odd.

$\square$

7.5 Compact Linear Operators

One of the most useful applications of the antipodal theorem is the fact, that the degree is different from zero. We will continue these considerations and specialize to linear compact operators.

Theorem 7.8 *The product formula for the Leray - Schauder degree: Let $\Omega \subset X$ be open bounded, $F_0 : \overline{\Omega} \rightarrow X$ compact, $G_0 : X \rightarrow X$ compact, $F = I - F_0, G = I - G_0, y \notin GF(\partial\Omega)$ and $K_\lambda, \lambda \in \Lambda$ the connected components of $X \setminus F(\partial\Omega)$. Then*

$$D(G, F, \Omega, y) = \sum_{\lambda \in \Lambda} D(F, \Omega, K_\lambda) D(G, K_\lambda, y)$$

where only finitely many terms are non zero and

$$D(F, \Omega, K_\lambda) = D(F, \Omega, z) \text{ for any } z \in K_\lambda.$$

This product formula in the sequel we need only for linear maps and $y = 0$, thus we will omit the proof of Theorem 7.8. Unfortunately, for the simplest proof of the product formula we need the approximation property of X, but in spite of this loss of generalization we will prove the following results.

Nonlinear Functional Analysis

Proposition 7.2 *Let $S, T : X \to X$ be compact linear operators, such that 1 is not an eigenvalue of S or of T. Then for $r > 0$*

$$D((I - S)(I - T), B(0, r), 0)$$
$$= D(I - S, B(0, r), 0).D(I - T, B(0, r), 0).$$

Proof :

Let $B = B(0, r)$. Since $0 \notin (I - S)(\partial B) \cup (I - T)(\partial B)$, all degrees are defined, and the mappings $I - S$, $I - T$ and $(I - S)(I - T)$ are isomorphisms, hence $\ker(I - S)$, $\ker(I - T)$, $\ker(I - S)(I - T)$ consist only of the zero element.

Let us now assume that X has the approximation property, i.e. every compact linear map is the uniform limit of compact linear maps with finite rank, we find finite rank operators S_0, T_0 such that $\|S - S_0\|$ and $\|T - T_0\|$ are sufficiently small. Then

$$
\begin{aligned}
D((I - S)(I - T), B, 0) &= D((I - S_0)(I - T_0)|_{\overline{B}_0}, B_0, 0) \\
&= sgn\ det(I - S_0)(I - T_0)|_{\overline{B}_0} \\
&= sgn\ det(I - S_0)|_{\overline{B}_0}.det(I - T_0)|_{\overline{B}_0} \\
&= D(I - S_0, B, 0).D(I - T_0, B, 0)
\end{aligned}
$$

$\square$

The following result is a special case of Proposition 7.2.

Theorem 7.9 *Let $X = X_1 \oplus X_2$ be a topological composition, and $T : X \to X$ a compact linear operator with $TX_j \subset X_j$, $j = 1, 2$. Let $I - T$ be an isomorphism of X. Then for each open ball $B = B(0, r)$ in X*

$$D(I - T, B, 0) = D(I - T|_{x_1}, B \cap X_1, 0).D(I - T|_{x_2}, B \cap X_2, 0).$$

The Topological Degree 101

Proof :

Let $B_j = B \cap X_j$, $P_j : X \to X_j$ be the linear projection onto X_j and $A_j = (I - T)P_j$. Then $x = P_1 x + P_2 x$. Let

$$S_1 = (I - T)P_1 + P_2, \quad S_2 = P_1 + (I - T)P_2.$$

S_j is injective, since $S_1 x = 0$ implies $P_2 x = 0$ and $(I - T)P_1 x = 0$ but $I - T$ is injective, hence $P_1 x = 0$. Observe that $S_j = I - TP_j$, and 1 is not an eigenvlaue of T, thus S_j is an isomorphism. $I - S_j = TP_j$ is compact, $(I - S_j)(X) \subset X_j$, by the reduction property (9) of the Leray - Schauder degree we obtain

$$
\begin{aligned}
D(S_j, B_1 + B_2, 0) &= D(S_j|_{(B_1+B_2)\cap X_j}, (B_1 + B_2) \cap X_j, 0) \\
&= D(S_j, B_j, 0).
\end{aligned}
$$

$$S_1 S_2 = (I - TP_1)(I - TP_2) = I - T(P_1 + P_2) + TP_1 TP_2 = I - T$$

since

$$TP_1 TP_2 = TP_1 P_2 T = 0$$

implies by Proposition 7.2

$$
\begin{aligned}
D(I - T, B_1 + B_2, 0) &= D(S_1 S_2, B_1 + B_2, 0) \\
&= D((I - TP_1)|_{X_1}, B_1, 0).D(I - TP_2)|_{X_2}, B_2, 0) \\
&= D(I - T|_{X_1}, B_1, 0).D(I - T|_{X_2}, B_2, 0).
\end{aligned}
$$

$\square$

Theorem 7.10 *Let X be a Banach space, $T : X \to X$ be linear and compact, let $\sigma(T) = \{\lambda \in \mathcal{C} : T - \lambda I$ is not continuously invertible $\}$ be the spectrum of T. Then*

102   Nonlinear Functional Analysis

1^0 $\sigma(T) \subset \{\lambda \in \mathcal{C} : |\lambda| \leq ||T||\}$
   $\sigma(T)$ *is countable, the only point of accumulation is*
   possibly 0 .

2^0 *If* $\lambda \notin \sigma(T), T - \lambda I$ *is an isomorphism in* X .

3^0 *If* $\lambda \in \sigma(T) \setminus \{0\}$, *there exists a minimal number*
   $k(\lambda) \in \mathcal{N}$, *such that for*
   $R(\lambda) = (\lambda I - T)^{k(\lambda)}(X), N(\lambda) = ker(\lambda I - T)^{k(\lambda)}$
   (a) $X = R(\lambda) \oplus N(\lambda)$
   (b) $TR(\lambda) \subset R(\lambda), TN(\lambda) \subset N(\lambda)$.

The number $n(\lambda) = \dim N(\lambda)$ is called the **algebraic multiplicity** of the eigenvalue λ, the **geometric multiplicity** is $\dim ker(T - \lambda I)$.

Theorem 7.11 *Let* X *be a real Banach space,* $T : X \to X$ *be a compact linear operator,* $\lambda \neq 0$ *and* λ^{-1} *not an eigenvalue of* T . *Let* $\Omega \subset X$ *be open and bounded and* $0 \in \Omega$. *Then* $D(I - \lambda T, \Omega, 0) = (-1)^{m(\lambda)}$ *where* $m(\lambda)$ *is the sum of the algebraic multiplicites of the eigenvalues* μ *of* T; *satisfying* $\mu\lambda > 1$, *and* $m(\lambda) = 0$ *if* T *has no eigenvalue of this kind.*

Proof :

Let $S = I - \lambda T = -\lambda(T - \lambda^{-1}I)$ is a homeomorphism onto X. Hence it is sufficient to consider $\Omega = B$, the unit ball. By theorem (7.10, (1^0)) there are at most finitely many eigenvalues $\mu \in \sigma(T)$ with $\lambda\mu > 1$, i.e. $sgn\ \mu = sgn\ \lambda$ and $|\mu| > |\lambda^{-1}|$, say $\mu_1, ..., \mu_p$. Let

$$V = \oplus_{j=1}^{p} N(\mu_j), \quad W = \cap_{j=1}^{p} R(\mu_j).$$

The Topological Degree

We show that $X = V \oplus W$. First of all $V \cap W = \phi$, since $x \in V \cap W$ implies

$$x = \sum_{j=1}^{p} x_j, \ x_j \in n(\mu_j) \text{ and } x \in R(\mu_j)$$

for $j = 1, 2, ..., p$. By theorem $(7.10, (3^0))$ we have

$$x_2 + x_3 + ... + x + p \in R(\mu_1)$$

hence

$$x_1 = x - \sum_{j=2}^{p} x_j \in R(\mu_1) \cap N(\mu_1) = \{0\}$$

and similarly we obtain $x_2 = ... = x_p = 0$. Now, any $x \in X$ may be written as $x = x_j + y_j$ with $x_j \in n(\mu_j), y_j \in R(\mu_j)$ by theorem $(7.10, (3^0))$ again, we have

$$x - \sum_{j=1}^{p} = x - x_k - \sum_{j \neq k} x_j = y_k - \sum_{j \neq k} x_j \in R(\mu_j)$$

hence

$$x - \sum_{j=1}^{p} x_j \in W = \cap R(\mu_k)$$

and

$$X = V \oplus W$$

By Theorem 7.9 we have

$$D(S, \Omega, 0) = D(S|_V, \Omega \cap V, 0).D(S|_W, \Omega \cap W, 0)$$

But

$$D(S|_W, \Omega \cap W, 0) = 1$$

since $T|_W$ has no eigenvalue μ with $\mu\lambda > 1$ and $x - t\lambda Tx$ defines an admissible homotopy from $I - \lambda T$ to I. By the same theorem we have

$$D(S|_V, \Omega \cap V, 0) = \prod_{j=1}^{p} d(S|_{N(\mu_j)}, \Omega \cap N(\mu_j), 0).$$

Since $h(t,x) = (2t-1)x - t\lambda Tx$ is an admissible homotopy from $S|_{N(\mu_j)}$ to $-I|_{N(\mu_j)}$ (this is true because $(2t-1)x - t\lambda Tx = 0$ and $||x|| = 1$ implies $\mu_j = \frac{2t-1}{t\lambda}$, hence $t = \frac{1}{2-\lambda\mu_j} > 1$) thus

$$D(S|_{N(\mu_j)}, \Omega \cap N(\mu_j), 0) = D(-I|_{N(\mu_j)}, \Omega \cap N(\mu), 0) = (-1)^{n(\mu_j)}$$

and therefore

$$D(S, \Omega, 0) = (-1)^{m(\lambda)},$$

where $m(\lambda) = \sum_{\lambda\mu_j > 1} n(\mu_j)$.

If there are no such μ at all, then $X = W$ and $D(S, \Omega, 0) = 1 = (-1)^0$.

$\square$

Chapter 8

BIFURCATION THEORY

8.1 An Example

Let X be a Banach space, $\Omega \subset X$ an open bounded subset, and $F : \overline{\Omega} \to X$ compact. We consider problems of the following type. Assume $0 \in \Omega$ and $F(0) = 0$. Then for every real λ the equation

$$\lambda x = F(x) \qquad (*)$$

has trivial solution. The following example shows that there exist real λ_0 that

$$\forall \epsilon > 0 \exists (\lambda, x_\lambda) \ 0 < ||x_\lambda|| < \epsilon, \ |\lambda - \lambda_0| < \epsilon$$

$$\lambda x_\lambda = F(x_\lambda),$$

i.e. a branch of nontrivial solutions of $(*)$ starts in the point $(\lambda_0, 0)$. The point $(\lambda - 0, 0) \in \mathcal{R} \times \Omega$ is called a bifurcation point. We will solve the following integral equation

$$\lambda x(s) = \frac{2}{n} \int_0^\pi [a \ sin \ s \ sin \ t + b \ sin \ 2s \ sin \ 2t][x(t) + x^3(t)]dt \tag{8.1}$$

105

106 Nonlinear Functional Analysis

which has a second-rank kernel. We suppose that $0 < b < a$. Because of the form of the kernel, any solution of equation (8.1) is necessarily of the form $x(s) = A \sin s + B \sin 2s$ with undetermined constants A, B (which will turn out to be functions of the real parameter λ). Substituting in equation (8.1), we have

$$\lambda[A \sin s + B \sin 2s] = \frac{2}{\pi} \int_0^\pi [a \sin s \sin t + b \sin 2s \sin 2t]$$
$$.[A \sin t + B \sin 2t + (A \sin t + B \sin 2t)^3]dt$$
$$= \frac{2}{n} a. \sin s[A \int_0^\pi \sin^2 t dt + A^3 \int_0^\pi \sin^4 t dt$$
$$+3AB^2 \int_0^\pi \sin^2 t. \sin^2 2t dt]$$
$$+\frac{2}{n} b. \sin 2s[B \int_0^\pi \sin^2 t dt + 3A^2 B \int_0^\pi \sin^2 2t. \sin^2 t dt$$
$$+B^3 \int_0^\pi \sin^4 2t dt]$$
$$= \frac{2}{\pi} a. \sin s[\frac{\pi}{2}A + \frac{3\pi}{8}A^3 + \frac{3\pi}{4}AB^2]$$
$$+\frac{2}{\pi} b. \sin 2s[\frac{\pi}{2}B + \frac{3\pi}{4}A^2 B + \frac{3\pi}{8}B^3],$$

where use has been made of the following values of integrals:

$$\int_0^\pi \sin t. \sin 2t dt = \int_0^\pi \sin^3 t. \sin 2t dt = \int_0^\pi \sin t. \sin^3 2t dt = 0.$$

Equating coefficients of $\sin s$ and $\sin 2s$, we obtain a pair of nonlinear simultaneous algebraic equations:

$$\lambda A = aA + \frac{3}{4}aA^3 + \frac{3}{2}aAB^2$$

$$\lambda B = bB + \frac{3}{2}bA^2 B + \frac{3}{4}bB^3. \tag{8.2}$$

There are four kinds of solutions of equations:

1. $A = B = 0$; this gives the trivial solution of equation (8.1).

Bifurcation Theory

2. $A \neq 0, B = 0$; only the first equation is nontrivial. We cancel $A \neq 0$ to obtain

$$\lambda = a + \frac{3}{4}aA^2$$

whence

$$A = \pm\frac{2}{\sqrt{3}}\sqrt{\lambda/a - 1}.$$

The corresponding solution of equation (8.1) is

$$x_1(s, \lambda) = \pm\frac{2}{\sqrt{3}}\sqrt{\lambda/a - 1}\sin s,$$

defined and real for $\lambda \geq a$.

3. $A = 0, B \neq 0$; only the second equation is nontrivial. We cancel $B \neq 0$ to obtain

$$\lambda = b + \frac{3}{4}bB^2$$

whence

$$B = \pm\frac{2}{\sqrt{3}}\sqrt{\lambda/b - 1}.$$

The corresponding solution of equation (8.1) is

$$x_2(s, \lambda) = \pm\frac{2}{\sqrt{3}}\sqrt{\lambda/b - 1}\sin 2s,$$

defined and real for $\lambda \geq b$, where we recall that $b < a$.

4. $A \neq 0, B \neq 0$; here both A and B may be cancelled in equation (8.2).

We obtain two ellipses:

$$\frac{3}{4}A^2 + \frac{3}{2}B^2 = \frac{\lambda}{a} - 1$$

$$\frac{3}{2}A^2 + \frac{3}{4}B^2 = \frac{\lambda}{b} - 1.$$

(8.3)

Solutions of equation (8.3) are given by intersections of these ellipses. Solving, we get

$$A^2 = \frac{4}{9} \cdot \left[\frac{2a-b}{ab} \lambda - 1 \right], \quad B^2 = \frac{4}{9} \cdot \left[\frac{2b-a}{ab} \lambda - 1 \right],$$

so that we have the following solutions of equation (8.1):

$$x_3(s, \lambda) = \pm \frac{2}{3} \cdot \sqrt{\frac{2a-b}{ab} \lambda - 1} \sin s \pm \frac{2}{3} \cdot \sqrt{\frac{2b-a}{ab} \lambda - 1} \sin 2s.$$

$$(8.4)$$

Clearly $2a - b > 0$ since we assumed that $b < a$. Hence the question of whether or not solutions of the form of Equation (8.4) can be real hinges upon whether or not $2b - a > 0$, or $\frac{b}{a} > \frac{1}{2}$.

We have the following cases:

Case I : $\frac{b}{a} < \frac{1}{2}$; $x_3(s, \lambda)$ is real for no real λ.

Case II : $\frac{b}{a} > \frac{1}{2}$; $x_3(s, \lambda)$ is real for $\lambda > \max(\frac{ab}{2a-b}, \frac{ab}{2b-a})$.

Since $a > b$, this means $x_3(s, \lambda)$ is real when $\lambda > \frac{ab}{2b-a}$.

In Case I above, i.e. when $\frac{b}{a} \le \frac{1}{2}$, the only real solutions of equation (8.1) are the trivial solution $x(s, \lambda) \equiv 0$, and the two main branches:

$$x_1(s, \lambda) = \pm \frac{2}{\sqrt{3}} \cdot \sqrt{\frac{\lambda}{a} - 1} \sin s$$

$$x_2(s, \lambda) = \pm \frac{2}{\sqrt{3}} \cdot \sqrt{\frac{\lambda}{b} - 1} \sin 2s$$

The solutions x_1 and x_2 branch away from the trivial solution $x \equiv 0$ at the eigenvalues a, b of the linearization of equation (8.1) at the origin:

$$\lambda h(s) = \frac{2}{\pi} \int_0^\pi [a \sin s \sin t + b \sin 2s \sin 2t] h(t) dt. \qquad (8.5)$$

Bifurcation Theory

In Case II, i.e. when $\frac{b}{a} > \frac{1}{2}$, we again have trivial solution $x(s, \lambda) \equiv 0$, and the two main branches

$$x_1(s, \lambda) = \pm \frac{2}{\sqrt{3}} \cdot \sqrt{\frac{\lambda}{a} - 1} \sin s$$

$$x_2(s, \lambda) = \pm \frac{2}{\sqrt{3}} \cdot \sqrt{\frac{\lambda}{b} - 1} \sin 2s$$

which bifurcate from $x \equiv 0$ at the primary bifurcation points, which are the eigenvalues a, b of linearized equation (8.5). Moreover, for $\lambda > \frac{ab}{2b-a} > a$, a third type of solution branch appears, namely that in equation (8.4). Note that as $\lambda \to \frac{ab}{2b-a}, \lambda > \frac{ab}{2b-a}$, the coefficients $\sqrt{\frac{2b-a}{ab}\lambda - 1} \to 0$ and $\sqrt{\frac{2b-a}{ab}\lambda - 1} \to \sqrt{\frac{3a-3b}{2b-a}}$. On the other hand note that $\sqrt{\frac{\lambda}{a} - 1} \to \sqrt{\frac{a-b}{2b-a}}$ as $\lambda \to \frac{ab}{2b-a}$. Thus as $\lambda \to \frac{ab}{2b-a}$, we see that $x_3(s, \lambda) \to x_3[s, \frac{ab}{2b-a}] = x_1[s, \frac{ab}{2b-a}]$. Therefore at $\lambda = \frac{ab}{2b-a}$, the sub-branch (twig)

$$x_3^+(s, \lambda) = \frac{2}{3}\sqrt{\frac{2b-a}{ab}\lambda - 1} \sin s \pm \frac{2}{3}\sqrt{\frac{2b-a}{ab}\lambda - 1} \sin 2s$$

joins the main branch, i.e. $x_3^+[s, \frac{ab}{2b-a}] = x_1^+[s, \frac{ab}{2b-a}]$ while the sub-branch (twig)

$$x_3^-(s, \lambda) = -\frac{2}{3}\sqrt{\frac{2b-a}{ab}\lambda - 1} \sin s \pm \frac{2}{3}\sqrt{\frac{2b-a}{ab}\lambda - 1} \sin 2s$$

joins the negative part of the main branch, i.e., $x_3^-[s, \frac{ab}{2b-a}] = x_1^-[s, \frac{ab}{2b-a}]$.

We have in Case II, when $\frac{b}{a} > \frac{1}{2}$, the phenomena of "secondary bifurcation", or the forming of sub-branches or twigs which bifurcate from the main branches. The main branches bifurcate from the trivial solution at the eigenvalues of the linearization of equation (8.5), while the twigs bifurcate from the main branches.

110 Nonlinear Functional Analysis

Thus solutions of the nonlinear equation (8.1) exist as continuous loci in $(\lambda, \sin s, \sin 2s)$ space. There are two main branches: $x_1(s, \lambda)$ splits off from the trivial solution $x \equiv 0$ at $\lambda = a$, and its two parts x_1^+, x_1^- differ only in sign; $x_2(s, \lambda)$ joins the trivial solution at $\lambda = b$, and its two parts x_2^+, x_2^- differ only in sign. a and b on the λ axis are the primary bifurcation points for the main branches. If $\frac{b}{a} > \frac{1}{2}$, i.e. Case II, two sub-branches or twigs split away from $x_1(s, \lambda)$ at $\lambda = \frac{ab}{2b-a}$, which is known as a secondary bifurcation point.

The question of whether or not secondary bifurcation of the eigensolutions of equation (8.1) takes place therefore hinges on whether we have $\frac{b}{a} > \frac{1}{2}$, or $\frac{b}{a} \leq \frac{1}{2}$. The condition $\frac{b}{a} \leq \frac{1}{2}$ in this simple problem is a "condition preventing secondary bifurcation".

8.2 Local Bifurcation

We will use the Leray - Schauder degree to study bifurcation points.

Definition 8.1 *Let X, Y be Banach spaces, $\Omega \subset X$ open, $0 \in \Omega$, $F : (a, b) \times \Omega \to Y$ continuous, such that for all $\lambda \in (a, b)$ we have $F(\lambda, 0) = 0$. The point $(\lambda_0, 0) \in (a, b) \times \Omega$ is said to be a bifurcation point, iff for all $\epsilon > 0$ there exist $x_\lambda \in \Omega$ and $\lambda \in (a, b)$ with $|\lambda - \lambda_0| < \epsilon, 0 < \|x_\lambda\| < \epsilon$, such that*

$$E(\lambda, x_\lambda) = 0.$$

If we assume that F can be linearized near $(\lambda_0, 0)$, then it is easy to give necessary conditions for bifurcation in terms of the linearization.

Bifurcation Theory

Proposition 8.1 *In the situation of Definition 8.1 let*

(a) F, F_x' *are continuous in a neighbourhood of the bifurcation point* $(\lambda_0, 0)$. *Then* $F_x'(\lambda_0, 0)$ *is not a homeomorphism.*

(b) *If* $X = Y, F(\lambda, x) = x - \lambda T x + G(\lambda, x)$ *with a continuous* $G : (a, b) \times \Omega \to X$, *such that*

$$\sup_{\lambda \in (a,b)} \frac{\|G(\lambda, x)\|}{\|x\|} \to 0$$

if $\|x\| \to 0$, *then* λ_0^{-1} *belongs to the spectrum of* T.

Proof :

(a) If $F_x'(\lambda_0, 0)$ is a homeomorphism then the implicit function Theorem 4.1 tells us, that F has a unique solution, i.e. only the trivial solution near $(\lambda_0, 0)$.

(b) If $\lambda_0^{-1} \notin \sigma(T)$, then $I - \lambda_0 T$ is an isomorphism for all λ close to λ_0 and

$$x = -(I - \lambda T)^{-1} G(\lambda, x) \neq 0$$

contradicts

$$1 \leq \|(I - \lambda T)^{-1}\| \cdot \frac{\|G(\lambda, x)\|}{\|x\|} \to 0.$$

$\square$

If $G = 0$, i.e. if F is linear, and λ_0^{-1} is an eigenvalue of T with an eigenvector x_0, then for all $\alpha \in \mathcal{R}$ the pair $\lambda(\alpha) = \lambda_0$, $x(\alpha) = \alpha x_0$ solves

$$F(\lambda(\alpha), x(\alpha)) = 0.$$

112 Nonlinear Functional Analysis

In this case $(\lambda_0, 0)$ is called a vertical bifurcation point.

Example

Let $X = Y = \mathcal{R}^2, x = (\xi, \eta), F(\lambda, x) = (1 - \lambda)(\xi, \eta) + (\eta^3, -\xi^3)$ then $F_x'(1,0) = 0$ but $F(\lambda, x) = 0$ implies

$$(1 - \lambda)\xi + \eta^3 = (1 - \lambda)\eta - \xi^3 = 0$$

and $(1 - \lambda)\xi\eta = -\eta^4 = \xi^4$, hence $\xi = \eta = 0$, i.e. $(1,0) \in \mathcal{R} \times X$ is not a bifurcation point.

If we represent F in the form of Proposition 8.1 (b), then $F(\lambda, x) = x - \lambda x + G(\lambda, x)$ with $G(\lambda, x) = (\eta^3, -\eta^3)$, thus $\lambda_0 = 1$ is of multiplicity two.

Example

Let $X = Y = \mathcal{R}^2$ and

$$F(\lambda, x) = \begin{pmatrix} \xi \\ \eta \end{pmatrix} - \lambda \begin{pmatrix} 0 & 1 \\ 1 & 1 \end{pmatrix} \begin{pmatrix} \xi \\ \eta \end{pmatrix} + \lambda \begin{pmatrix} 0 \\ \xi^3 \end{pmatrix}.$$

$\lambda_0 = 1$ has geometric multiplicity one, algebraic multiplicity two, but $F(\lambda, x) = 0$ implies

$$\xi - \lambda\eta = 0$$
$$\eta - \lambda(\xi + \eta) + \lambda\xi^3 = 0$$

hence

$$\eta(1 - \lambda^2 - \lambda + \lambda^4\eta^2) = 0,$$

therefore $\eta = 0$ or $\eta^2 = -\lambda^{-4}(1 - \lambda - \lambda^2)$, but the second solution is not close to the trivial solution, hence $(1, 0)$ is not a bifurcation point.

The following theorem will be based of the following degree jump principle: Let

$$x - H(\mu, x) = 0 \qquad (*)$$

Bifurcation Theory

$\mu \in \mathcal{R}, x \in X$ and assume for $\Omega \subset \mathcal{R} \times X$ bounded and open

(JP1) $H : \Omega \to X$ is compact and $H(\mu, 0) = 0$ for all μ

(JP2) For $\mu_1 < \mu_2$ we have

$$D(I - H(\mu_1, .), X \cap \Omega, 0) \neq D(IO - H(\mu_2, .), X \cap \Omega, 0).$$

Now if (JP1) is satisfied and $(\mu_0, 0)$ is not a bifurcation point of $(*)$, then $D(I - H(\mu, .), X \cap \Omega, 0)$ is constant in a neighbourhood of μ_0. If (JP1) and (JP2) are satisfied, then $(*)$ has a bifurcation point $(\mu, 0)$ with $\mu_1 < \mu < \mu_2$.

In the first case $(\mu, 0)$ is the only solution of $(*)$, hence homotopy invariance yields the constancy of the degree; in the second case the jump of the degree yields the existence of a nontrivial solution $(\mu, x) \neq (\mu, 0)$.

We will now assume that λ_0^{-1} is an eigenvalue of odd algebraic multiplicity.

Theorem 8.1 *Let X be a real Banach space, $K : X \to X$ be compact and linear, $\Omega \subset \mathcal{R} \times X$ a neighbourhood of $(\lambda_0, 0), G : \Omega \to X$ be compact and $G(\lambda, 0) = 0$. Suppose also that*

(a) λ_0^{-1} is an eigenvalue of K of odd algebraic multiplicity.

(b) There exists a continuous function $\varphi : \mathcal{R} \to \mathcal{R}$ with $\lim_{r \to 0} \varphi(r) = 0$ and $\delta > 0$, such that for all $(\lambda, x)(\lambda, \overline{x}) \in \Omega$ with $|\lambda - \lambda_0| \leq \delta, \|x\| \leq r, \|\overline{x}\| \leq r$

$$\|G(\lambda, x) - G(\lambda, \overline{x})\| \leq \varphi(r).\|x - \overline{x}\|.$$

Then $(\lambda_0, 0)$ is a bifurcation point for

$$F(\lambda, x) = x - \lambda K x + G(\lambda, x) = 0.$$

Nonlinear Functional Analysis

Proof :

1. Let $T = I - \lambda_0 K$, and $k = k(\lambda)$ be the smallest number, such that $N(\lambda_0) = \ker T^k = \ker T^{k+1}$. Then, according to Theorem 7.2 $X = N(\lambda_0) \oplus R(\lambda_0)$, and both subspaces are invariant under $K.T$ is a homeomorphism from $R(\lambda_0)$ onto itself and λ_0^{-1} is the only eigenvalue of $K|_{N(\lambda_0)}$. Let $P : X \to N(\lambda_0)$ be the projection defined by the direct sum. Then Range $(I - P) = R(\lambda_0)$ and writing $v = Px, z = (I - P)x$, we have $F(\lambda_0 + \mu, x) = 0$ equivalent to

$$x = v + z$$

$$\lambda = \lambda_0 + \mu$$

$$Tv = \mu K v - P\overline{G}(\mu, v + z) \tag{8.1}$$

$$Tz = \mu K z - (I - P)\overline{G}(\mu, v + z) \tag{8.2}$$

where $\overline{G}(\mu, x) = G(\lambda_0 + \mu, x)$.

Let $S = (T|_{R(\lambda_0)})^{-1}$. Then the second equation becomes

$$z = \mu SK z - S(I - P)\overline{G}(\mu, v + z). \tag{8.3}$$

2. We will solve equation (8.3) by applying Banach's fixed point theorem. If $|\mu| \leq \eta, ||v|| \leq r$ are sufficiently small, then G is a contraction in the space of $R(\lambda_0)$ - valued continuous functions. Let

$$J = [-\eta, \eta], B = \overline{B}(0, r) \cap N(\lambda_0), \text{ and for } \rho, c > 0$$

$$M = \{z : J \times b \to R(\lambda_0) \text{ continuous with}$$

$$||z(\mu, v)|| \leq c||v||, \sup_{\mu, v} ||z(\mu, v)|| \leq \rho\}.$$

Then for $z, z_1, z_2 \in M, \mu \in J$

$$||\overline{G}(\mu, v + z_1) - \overline{G}(\mu, v + z_2)|| \, eq\varphi(r)||z_1 - z_2||$$

$$||\overline{G}(\mu, v + z)|| = ||\overline{G}(\mu, v + z) - \overline{G}(\mu, 0)||$$

$$\leq \varphi(r)||v + z|| \leq \varphi(r)(1 + c)||v|| \leq \varphi(r)(1 + c)r.$$

Bifurcation Theory 115

The operator

$$z \to (I - \mu SK)^{-1} S(I - P)\overline{G}(\mu, v + z)$$

maps M into itself and is a contraction, if r (and thus $\varphi(r)$) is chosen sufficiently small.

Thus there exist a unique $z(\mu, v)$, which solves equation (8.3), and $w(\mu, v) = ||v||^{-1} z(\mu, v)$ satisfies $||w||_\infty \leq c$. Hence equation (8.2) implies by

$$||w(\mu, v)|| \leq ||(I - \mu SK)^{-1} S(I - P)||.||\overline{G}(\mu, v + ||v||w(\mu, v))||.||v||^{-1}$$

uniformly in $\mu \in J$

$$\lim_{||v|| \to 0} ||w(\mu, v)|| = 0.$$

Thus, all possible zeros of F in a small neighbourhood of $(\lambda_0, 0)$ are contained in

$$\{(\lambda, x) : \lambda = \lambda_0 + \mu, x = v + mz(\mu, v), |\mu| \leq \eta, ||v|| \leq r\}.$$

3. Now, let us insert $z = z(\mu, v)$ into the bifurcation equation (8.1) on $N(\lambda_0)$. Let $G_0 : J \times B \to n(\lambda_0)$ be defined by

$$G_0(\mu, v) = (I - (\lambda_0 + \mu)K)v + P\overline{G}(\mu, v + z(\mu, v)).$$

Clearly, G_0 is continuous, and is of the form $I - F_0$ with a compact map F_0. Now we will apply degree theory.

Choose μ_1, μ_2 such that $-\eta \leq \mu_1 < 0 < \mu_2 \leq \eta$ and $\rho \in (0, r]$ such that for $j = 1, 2$ we have in $N(\lambda_0) \setminus \{0\}$

$$G_0(\mu_1, .) \text{ is homotopic to } I - (\lambda_0 + \mu_1)K|_{N(\lambda_0)}$$

$$G_0(\mu_2, .) \text{ is homotopic to } I - (\lambda_0 + \mu_2)K|_{N(\lambda_0)}$$

in $N(\lambda_0) \setminus \{0\}$, e.g.

$$H(t, \mu, v) = (I + (\lambda_0 + \mu)K)v + tP\overline{G}(\mu, v + z)$$

is an admissible homotopy, since $H(t, \mu, v) = 0$ for $\mu \neq 0$ implies

$$v = -t(I + (\lambda_0 + \mu)K)^{-1}P\overline{G}(\mu, v + z)$$
$$1 \leq t||(I + (\lambda_0 + \mu)K)^{-1}P||.|\overline{G}(\mu, v + z)||.||v||^{-1} \to 0$$

Then

$$D(G_0(\mu_j, .), B(0, \rho), 0) = D(I - (\lambda_0 + \mu_j)K, B(0, \rho), 0).$$

By Theorem 7.11

$$D(I - (\lambda_0 + \mu_j)K, B(0, \rho), 0) = (-1)^{m(\lambda_0 + \mu_j)} \qquad (8.4)$$

where $m(\lambda_0 + \mu_j)$ is the sum of the algebraic multiplicities of the eigenvalues λ of $K|_{N(\lambda_0)}$ satisfying $\lambda(\lambda_0 + \mu_j) > 1$, or 0, if there are no such eigenvalues. Since λ_0^{-1} is the only eigenvalue of $K|_{N(\lambda_0)}$ and either $\lambda_0^{-1}(\lambda_0 + \mu_1) > 1$ and $\lambda_0^{-1}(\lambda_0 + \mu_2) < 1$ or vice versa, one of the degrees in equation (8.4) is $+ 1$ while the other one is -1.

Hence G_0 must have a zero in $[\mu_1, \mu_2] \times \partial B(0, \rho)$ since G_0 would be an admissible homotopy otherwise. If (μ_ρ, v_ρ) is such a zero, then $(\lambda_0 + \mu_\rho, x_\rho)$ with $x_\rho = v_0 + z(\mu_\rho, v_\rho)$ is a nontrivial zero of F. Since μ_j and ρ may be chosen arbitrarily close to zero, we have shown, that $(\lambda_0, 0)$ is a bifurcation point of $F(\lambda, x) = 0$. $\qquad \square$

8.3 Bifurcation and Stability

In this section we will study the situation of Theorem 8.3 more carefully. Again let us assume, that

$$F(\lambda, x) = (I - \lambda K)x + G(\lambda, x) \qquad (8.1)$$

Bifurcation Theory   117

where $K : X \to X$ is linear and compact, λ_0^{-1} is a simple eigenvalue of $K, G : \mathcal{R} \times \Omega$ is continuous differentiable and Ω is an open neighbourhood of 0, such that for all $\lambda \in \mathcal{R}$

(a) $G(\lambda, 0) = 0$

(b) $G'_x(\lambda, 0) = 0$

(c) $G(\lambda, .)$ is compact

(d) $\{G(., x), x \in \Omega\}$ is equicontinuous

(e) $\sigma(K) \setminus \{\lambda_0^{-1}\} \subset \{\zeta \in \mathcal{C} : Re\zeta < \lambda_0^{-1}\}$.

Let $x_0 \in X, \|x_0\| = 1$ be an eigenvector of K to the eigenvalue λ_0^{-1}, i.e. $(I - \lambda_0 K)x_0 = 0$ and let $x_0^* \in X^*$ be chosen, such that projection $P : X \to n(\lambda_0)$ is given by $Px = <x, x_0^*> x_0$.

Again we have the decomposition of $X = R(\lambda_0) \oplus N(\lambda_0)$ and equation (8.1) decomposes into

$$(I - \lambda K)Px + PG(\lambda, x) = 0 \tag{8.2}$$

$$(I - \lambda K)(-P)x + (I - P)G(\lambda, x) = 0. \tag{8.3}$$

If we again denote by $T = I - \lambda_0 K$,

$$\lambda = \lambda_0 + \mu, \overline{G}(\mu, x) = G(\lambda_0 + \mu, x),$$

then by definition of P equation (8.2) reduces to

$$-\frac{\mu}{\lambda_0}\alpha\alpha_0 + P\overline{G}(\mu, x) = 0 \tag{8.4}$$

with $\alpha = <x, x_0^*>$. If we again denote by

$$S = [(I - \lambda_0 K)|_{R(\lambda_0)}]^{-1},$$

the equation (8.3) becomes

$$(I - P)x - \mu SK(I - P)\overline{G}(\mu, x) = 0$$

and with

$$H((\alpha, \mu), x) = x - \mu SK(I - P)x + S(I - P)\overline{G}(\mu, x) - \alpha\alpha_0$$

$$H((\alpha, \mu), x) = 0 \tag{8.5}$$

where $H : \mathcal{R}^2 \times \Omega \to X$ is continuously differentiable.

Since $H'_x(0, 0) = I$, by the implicit function theorem there exist neighbourhoods of 0 in $\mathcal{R}^2$ and in Ω and a continuously differentiable function

$$(\alpha, \mu) \to x(\alpha, \mu)$$

such that locally

$$H((\alpha, \mu), x(\alpha, \mu)) = 0, \quad x(0, 0) = 0.$$

Now we additionally assume

(f) There exists a continuously differentiable mapping $G_1 :$ $\mathcal{R} \times \Omega \times \mathcal{R} \to X$, such that for all $\mu, \alpha \in \mathcal{R}, x \in \Omega$ with $\alpha = <x, x_0^*>$

$$G(\lambda, \alpha x) = \alpha^2 G_1(\lambda, x, \alpha).$$

Equation (8.4) then has the form

$$-\frac{\mu}{\lambda_0} + \alpha < \overline{G}_1(\mu, x(\alpha, \mu), \alpha), x_0^* >= 0. \tag{8.6}$$

The function

$$f(\alpha, \mu) = -\frac{\mu}{\lambda_0} + \alpha < \overline{G}_1(\mu, x(\alpha, \mu), \alpha), x_0^* >$$

has the property

$$f(0, 0) = 0$$
$$f'_\mu(0, 0) = -\frac{1}{\lambda_0} \neq 0.$$

Bifurcation Theory 119

By the implicit function theorem there exists a continuously differentiable function $\mu : (-\alpha_0, \alpha_0) \to \mathcal{R}$, such that for all α with $|\alpha| < \alpha_0 \neq 0$

$$F(\alpha, \mu(\alpha)) = 0, \quad \mu(0) = 0$$

and so we obtain a solution $x(\alpha, \mu(\alpha))$ of equation (8.1).

Now in the sequel we will assume, that for sufficiently small $|\alpha| \neq 0$ we have $\mu'(\alpha) \neq 0$. In this case we have three possibilities for the behaviour of $x(\alpha, \mu(\alpha))$:

$1^0 \quad \mu'(\alpha) < 0$ for $\alpha < 0$ and $\mu'(\alpha) > 0$ for $\alpha > 0$.

$2^0 \quad \mu'(\alpha)$ does not change its sign for small $|\alpha| \neq 0$.

$3^0 \quad \mu'(\alpha) > 0$ for $\alpha < 0$ and $\mu'(\alpha) < 0$ for $\alpha > 0$.

In Case 1^0 for all $\alpha \neq 0, \mu(\alpha) > \mu_0$. In Case 3^0 $\mu(\alpha) < \mu_0$ for $\alpha \neq 0$.

If $\mu'(\alpha) \neq 0, \alpha \neq 0$, then λ_0 is not a vertical bifurcation point, since $\mu(\alpha) \neq \mu(0) = 0$.

Let (λ_0, x_0) be an isolated zero of F, and Ω an open neighbourhood of x_0, such that $\overline{\Omega}$ does not contain an additional solution of $F(\lambda, x) = 0$. Then

$$D(F, \Omega, 0) = i(I - F, \Omega)$$

is called the **fixed point index**.

We now will determine the index of the trivial solution:

The linear operator $I - \lambda K$ does not have negative eigenvalues if $\lambda < \lambda_0$, since

$$(I - \lambda K)x = \nu x \text{ implies } (1 - \nu)x = \lambda K x.$$

If $\nu < 0$, then $\frac{1-\nu}{\lambda} > \frac{1}{\lambda} > \frac{1}{\lambda_0}$ would be an eigenvalue of K. Then, since $G(\lambda, 0) = 0$, we have with $\Omega = B(0, \epsilon)$

$$i(I - F, \Omega) = i(\lambda K + G(\lambda, .), \Omega) = i(\lambda K, \Omega)$$

$$= D(I - \lambda K, \Omega, 0) = (-1)^{m(\lambda)} = 1$$

since $m(\lambda) = 0$ (no eigenvalues $> \lambda_0^{-1}$). If we choose $\lambda > \lambda_0$ such that λ_0^{-1} is the only eigenvalue of K in the set $\{\zeta \in \mathcal{C}, Re\zeta > \frac{1}{\lambda}\}$, then

$$i(I - F, \Omega) = i(\lambda K, \Omega) = (-1)^{m(\lambda)} = -1,$$

since $m(\lambda) = 1(\lambda_0^{-1}$ is a simple eigenvalue).

Thus the trivial solution has the index

$$i(\lambda K, \Omega) = 1 \text{ if } \lambda < \lambda_0$$

$$i(\lambda K, \Omega) = -1 \text{ if } \lambda < \lambda_0.$$

The index of the nontrivial solutions can be obtained by the following:

Since $\mu'(\alpha) \neq 0$ for small $\alpha \neq 0$, there is an $\epsilon > 0$ such that for all x with $0 < ||x|| < \epsilon$

$$(I - \lambda_0 K)x + G(\lambda_0, x) \neq 0.$$

Then there is $\delta > 0$ such that for all λ with $|\lambda - \lambda_0| \leq \delta, ||x|| = \epsilon$ we have

$$(I - \lambda K)x + G(\lambda, x) \neq 0 \qquad (8.7)$$

otherwise then we would have sequences $(\lambda_n), (x_n)$, such that $|\lambda - \lambda_0| \leq \frac{1}{n}, ||x_n|| = \epsilon$ and

$$0 = F(\lambda_n, x_n) = (I - \lambda_n K)x_n + G(\lambda_n, x_n)$$

$$(I - \lambda_0 K)x_n + G(\lambda_0, x_n) = (I - \lambda_n K)x_n + G(\lambda_n, x_n)$$

$$+(\lambda_n - \lambda_0)Kx_n + (G(\lambda_0, x_n) - G(\lambda_n, x_n)).$$

$\{Kx_n, n \in \mathcal{N}\}$ is bounded, $(\lambda_n - \lambda_0)Kx_n \to 0, G$ is equicontinuous, hence $G(\lambda_0, x_n) - G(\lambda_n, x_n) \to 0$. From this follows

$$\lim_{n \to \infty} (I - \lambda_0 K)x_n + G(\lambda_0, x_n) = 0.$$

Bifurcation Theory

Since $F(\lambda_0, .)$ is proper, there is an $\hat{x}, ||\hat{x}|| = \epsilon$ and $F(\lambda_0, \hat{x}) = 0$.

By equation (8.7)

$$H(t, x) = F(\lambda_0 + \delta(2t - 1), x) \neq 0 \text{ for } 0 \leq t \leq 1, ||x|| = \epsilon$$

thus $D(H(\lambda, .), B(0, \epsilon), 0) =: d(\lambda)$ is constant if $|\lambda - \lambda_0| \leq \delta$. If $\lambda \neq \lambda_0$, the solutions are isolated, therefore $d(\lambda)$ is the sum of indices of the trivial and the nontrivial solutions.

Case 1^0: For $\lambda < \lambda_0$, the trivial solution has index 1, hence $d(\lambda) = 1$, if $\lambda > \lambda_0$ the trivial solution has index - 1, hence the both nontrivial solutions for $\alpha < 0$ and $\alpha > 0$ have the index $+1$.

Case 2^0 : Since $\mu'(\alpha) \neq 0$, we have nontrivial solutions for $\lambda < \lambda_0$ and $\lambda > \lambda_0$, and the index of the nontrivial solution of absolute value 1, we have $d(\lambda) = 1 + $ index of nontrivial solution, if $\lambda < \lambda_0$ $d(\lambda) = -1 + $ index of nontrivial solution, if $\lambda > \lambda_0$ therefore, if $\lambda < \lambda_0$, then the index of the nontrivial and the index of the nontrivial solution is $+1$, if $\lambda > \lambda_0$, and $d(\lambda) = 0$.

Case 3^0 : $d(\lambda) = -1$, since the trivial solution has the index -1, if $\lambda > \lambda_0$, thus both nontrivial solutions have index -1.

When we have an evolution problem governed, for example, by the differential equation

$$x' = f(\lambda, x)$$

in an appropriate space, then the results tell us something about the existence and number of equilibria, i.e. time-independent solutions. Consider

$$x'(t) = f(\lambda, x(t)) \text{ with } f(\lambda, 0) = 0$$

and $f(\lambda_0, x) = Ax + R(x)$, where A is an $x \times n$ - matrix and $Rx = 0(||x||)$. Then the trivial solution $x = 0$ of $x' = f(\lambda_0, x), x(0) = x_0$ is said to be **stable**, if for every $\epsilon > 0$ there exists $\delta > 0$, such that the solution exists on $\mathcal{R}^+$ and satisfies $||x(t)|| \leq \epsilon$ whenever $||x_0|| \leq \delta$. The trivial solution of $x' = Ax$ is stable iff Re $\mu \leq 0$ for all $\mu \in \sigma(A)$ and every μ, such that Re $\mu = 0$ has algebraic multiplicity equal to its geometric one. If Re $\mu < 0$ for all $\mu \in \sigma(A)$ then $x = 0$ is stable as a solution of $x' = f(x)$. If Re $\mu > 0$ for some $\mu \in \sigma(A)$ then $x = 0$ is an unstable solution of $x' = f(x)$.

In our formulation the operator A is given by $-(I - \lambda K)$, and the condition (e) in section 8.3 states that 0 is a simple eigenvalue of A and all other eigenvalues of A have negative real part.

Since 0 is an isolated point of the specturm, we find $\rho > 0$, such that

$$(\sigma(I - \lambda_0 K) \setminus \{0\}) \cap B(0, 4\rho) = 0$$

and we also find $\eta > 0$ such that for all $x \in X, \lambda \in C$ with $||x|| < \eta, |\lambda - \lambda_0| < \eta$ implies

$$\sigma(I - \lambda K + G'_x(\lambda, x)) \subset \cup_{\zeta \in \sigma(I - \lambda_0 K)} B(\zeta, \frac{\rho}{2})$$

and $\{\zeta \in C : |\zeta| < \eta\} \cap \{\sigma(I - \lambda K + G'_x(\lambda, x))\} = \{\hat{\lambda}\}$, i.e. the λ - neighbourhood of the simple eigenvalue 0 of $I - \lambda_0 K$ contains only the simple eigenvalue $\hat{\lambda}$ of $I - \lambda K + G'_x(\lambda, x)$.

Since the spectral values are real or conjugate complex, $\hat{\lambda}$ is real; and $\hat{\lambda} \neq 0$ for small solutions $x = x(\alpha.\mu(\alpha))$, since $\mu'(\alpha) \neq 0$. Therefore the stability depends upon $\hat{\lambda} > 0$ or $\hat{\lambda} < 0$. Since λ_0 is not a vertical bifurcation point, for the nontrivial solution (λ, x) we have $\lambda \neq \lambda_0$.

Bifurcation Theory

For $\lambda > \lambda_0$ the index of the nontrivial solution is $+1$, and this is true if $\hat{\lambda} > 0$, these solutions are stable. If $\lambda < \lambda_0$, the index is -1, which requires $\hat{\lambda} < 0$, these solutions are unstable.

8.4 Global bifurcation

So far we have only considered local results, i.e. existence of solutions in small neighbourhoods of a bifurcation point. However, essentially under the same hypothesis, it is possible to prove more about the global behaviour of components of the solution set containing these points, as we are going to indicate in this section.

Let X be a real Banach space, $\Omega \subset \mathcal{R} \times X$ a neighbourhood of $(\lambda_0, 0), K$ be linear, $G : \overline{\Omega} \to X$ be continuous and such that $G(\lambda, x) = 0(\|x\|)$ as $x \to 0$, uniformly in λ. Typical global results about zeros of $F(\lambda, x) = x - \lambda K x + G(\lambda, x)$ will be explained for compact K and G. So, let λ_0^{-1} be an eigenvalue of odd algebraic multiplicity of the compact K. Let

$$M = \{(\lambda, x) \in \Omega : F(\lambda, x) = 0 \text{ and } x \neq 0\}$$

and C be the connected component of $\overline{M}$ containing $(\lambda_0, 0)$. Remember that components are closed and $(\lambda_0) \in \overline{M}$ since $(\lambda_0, 0)$ is a bifurcation point. We want to prove that $C \cap \partial \Omega \neq 0$ of $(\lambda_1, 0) \in C$ for another characteristic value $\lambda_1 \neq \lambda_0$ of K. In case $\Omega = \mathcal{R} \times X, C \cap \partial \Omega \neq 0$ means that C is unbounded.

So, let us first sketch how we arrive at a contradiction if we assume

$$C \cap \partial \Omega = 0, \quad C \cap (\mathcal{R} \times \{0\}) = \{(\lambda_0, 0)\}. \tag{8.1}$$

First of all $C \cap \partial \Omega = 0$ implies that C is compact, since K and G are compact. Suppose next that we are able to find an open bounded Ω_0 such that $C \subset \Omega_0 \subset \overline{\Omega}_0 \subset \Omega$ and $\overline{M} \cap \partial\Omega_0 = 0$. By the second part of equation (8.1) we may then assume that the intersection of $\overline{\Omega}_0$ and real line is given by $J = [\lambda_0 - \delta, \lambda_0 + \delta]$ with $\delta > 0$ so small that no other characteristic value of K satisfies $|\lambda - \lambda_0| \leq 2\delta$. By the homotopy invariance of the Leray - Schauder degree $\overline{M} \cap \partial\Omega_0 = 0$ then implies that $D(\lambda) = D(F(\lambda, .), \Omega_0(\lambda), 0)$ is constant in j; remember that $\Omega_0(\lambda) = \{x : (\lambda, x) \in \Omega_0\}$. To see this let $\alpha = D(F(\lambda_0, .), \Omega_0(\lambda_0), 0)$ and let

$$\hat{\lambda} = \inf\{\lambda, |\lambda - \lambda_0| \leq \delta, D(F(\lambda, .), \Omega(\lambda), 0) \neq \alpha\}$$

then there is an $\hat{x} \in \partial\Omega(\hat{\lambda})$, such that $(\hat{\lambda}, \hat{x}) \in \overline{M} \cap \partial\Omega_0$.

Like in the proof of Theorem 8.1, we want to exploit the jump in the degree when λ crosses λ_0. Hence, choose λ_1 and λ_2 such that $\lambda_0 - \delta < \lambda_1 < \lambda_0 < \lambda_2 < \lambda_0 + \delta$ and note that

$$D(\lambda_i) = D(F(\lambda_i, .), \Omega_0(\lambda_i) \setminus \overline{B}_\rho(0), 0) - D(F(\lambda_i, .), B_\rho(0), 0) \tag{8.2}$$

for $i = 1, 2$ with $\rho > 0$ sufficiently small. Since the $D(F(\lambda_i, .), B_\rho(0), 0)$ differ by a factor -1 and $D(\lambda_1) = D(\lambda_2)$, the first degrees on the right-hand side of equation (8.2) must also be different. But it is easy to see that they are in fact equal to zero. Indeed, consider for example $\lambda_3 > \lambda_2$ so large that $\Omega_0(\lambda_3) = 0$ and $\rho > 0$ so small that $F(\lambda, x) \neq 0$ on $\overline{B}_\rho(0) \setminus \{0\}$ for $\lambda \in [\lambda_2, \lambda_0 + 2\delta]$ and $\overline{\Omega}_0(\lambda) \cap \overline{B}_\rho(0) = 0$ for $\lambda \geq \lambda_0 + 2\delta$. Then the homotopy invariance for $\Omega_0 \setminus ([\lambda_2, \lambda_3] \times \overline{B}_\rho(0))$ implies

$$D(F(\lambda_2, .), \Omega_0(\lambda_2) \setminus \overline{B}_e(0), 0) = D(F(\lambda_3, .), \Omega(\lambda_3), 0) = 0.$$

Thus, the only problem is to find such a bounded nieghbourhood Ω_0 of C. Let us start with $U_\delta = \{(\lambda, x) \in \Omega : \text{dist}((\lambda, x), C) < \delta\}$. Evidently, $\overline{U}_\delta \cap \overline{M}$ is

Bifurcation Theory

compact and $C \cap \partial U_\delta = 0$. Note that $\overline{U}_\delta \cap \overline{M}$ is not connected unless it equals C, since C is already a maximal connected subset of $\overline{M}$. Of course we choose $\Omega_0 = U_\delta$ if $\overline{U}_\delta \cap \overline{M} = C$. If not then one may guess that, due to the disconnectedness of $\overline{U}_\delta \cap \overline{M}$, there exist compact $C_1 \supset C$ and $C_2 \supset \overline{M} \cap \partial U_\delta$ such that $C_1 \cap C_2 = \emptyset$ and $\overline{U}_\delta \cap \overline{M} = C_1 \cup C_2$. If this is true then
dist $(C_1, C_2) = \beta > 0$ and we may choose the intersection of U_δ and the $\beta/2$ - neighbourhood of C_1 for Ω_0.

Lemma 8.1 *Let (M, d) be a compact metric space. $A \subset M$ be a component and $B \subset M$ closed such that $A \cap B = 0$. Then there exist compact $M_1 \supset A$ and $M_2 \supset B$ such that $M = M_1 \cup M_2$ and $M_1 \cap M_2 = \emptyset$.*

Proof :

To use a good substitute for possibly missing pathwise connectedness, namely ϵ - chains, let us recall that, given $\epsilon > 0$, two points $a \in M$ and $b \in M$ are said to be $\epsilon - chainable$ if there are finitely many points $x_1, ..., x_n \in M$ such that $x_1 = a, x_n = b$ and $d(x_{i+1}, x_i) < \epsilon$ for $i = 1, ..., n - 1$. In this case $x_1, ..., x_n$ is an ϵ - chain joining a and b. Let

$A_\epsilon = \{x \in M : \text{there exists } a \in A \text{ such that } x \text{ and } a \text{ are } \epsilon \text{ -chainable} \}.$

Clearly $A \subset A_\epsilon$ and A_ϵ is both open and closed in M since

$B_\epsilon(z) \cap (M \setminus A_\epsilon) = \emptyset$ for $z \in A_\epsilon, B_\epsilon(z) \cap A_\epsilon = \emptyset$ for $z \in M \setminus A_\epsilon$.

It is therefore enough to show $B \cap A_\epsilon = \emptyset$ for some $\epsilon > 0$, since then $M_1 = A_\epsilon$ and $M_2 = M \setminus A_\epsilon$ have all properties we are looking for. Suppose, on the contrary, that $B \cap A_\epsilon \neq \emptyset$ for all $\epsilon > 0$. Consider $\epsilon_n \to 0, (a_n) \subset A$ and $(b_n) \subset B$ such that a_n and b_n are ϵ_n - chainable. Since A and B are compact, we may assume $a_n \to a_0 \in A$ and $b_n \to b_0 \in B$, and therefore we have

ϵ_n - chains M_n joining a_0 and b_0, for every $n \geq 1$. Consider the limit set

$$M_0 = \{x \in M : x = \lim_{k \to \infty} x_{n_k} \text{ with } x_{n_k} \in M_{n_k}\}.$$

Evidently, M_0 is compact and $a_0, b_0 \in M_0$. Suppose that M_0 is not connected. Then $M_0 = C_1 \cup C_2$ with C_i compact and dist $(U_\rho(C_1), U_\rho(C_2)) > \rho$ for sufficiently small $\rho > 0$. For $\epsilon_n < \rho$ this contradicts the obvious fact that any two $c_1 \in C_1$ and $c_2 \in C_2$ are ϵ_n - chainable. Hence, M_0 is connected. Consequently, $M_0 \subset A$ since $a_0 \in M_0 \cap A$ and A is maximal connected, and therefore $b_0 \in A \cap B$, a contradiction.

$$\square$$

The reasoning given so far leads to a further result, which we are going to prove next.

Theorem 8.2 *Let X be a real Banach space, $\Omega \subset \mathcal{R} \times X$ a neighbourhood of $(\lambda_0, 0), G : \overline{\Omega} \to X$ be completely continuous and $G(\lambda, x) = 0(\|x\|)$ as $x \to 0$, uniformly in λ. Let k be linear and compact and λ_0 a characteristic value of odd algebraic multiplicity $F(\lambda, x) = x - \lambda Kx + G(\lambda, x)$ and*

$$M = \{(\lambda, x) \in \Omega : F(\lambda, x) = 0 \text{ and } x \neq 0\}.$$

Then the component C of $\overline{M}$, containing $(\lambda_0, 0)$, has at least one of the following properties:

(a) $C \cap \partial\Omega \neq \emptyset$;

(b) C contains an odd number of trivial zeros $(\lambda_i, 0) \neq (\lambda_0, 0)$, where λ_1 is a characteristic value of K of odd algebraic multiplicity.

Exercises

Proof :

Suppose that $C \cap \partial\Omega = \emptyset$. Then we already know that C is compact and contains another $(\lambda, 0)$ with $\lambda \neq \lambda_0$. Clearly, a bounded neighbourhood Ω_0 of C satisfying $\overline{M} \cap \partial\Omega_0 = \emptyset$ contains only a finite number of points $(\lambda_k, 0)$ with $\lambda_k^{-1} \in \sigma(K)$, say $\lambda_1 < \ldots < \lambda_{i-1} < \lambda_0 < \lambda_{i+1} < \ldots < \lambda_p$. We may assume that $\overline{\Omega}_0 \cap (\mathcal{R} \times \{0\}) = \cup_{k=1}^p [\lambda_k - \delta, \lambda_k + \delta]$ with $\delta > 0$ sufficiently small. Choosing λ_{k1} and λ_{k2} such that $\lambda_k - \delta < \lambda_{k1} < \lambda_k < \lambda_{k2} < \lambda_k + \delta$, we have

$$D(F(\lambda, .), \Omega_0(\lambda), 0) = m \text{ on } [\lambda_1 - \delta, \lambda_p + \delta] \text{ for some } m \in \Pi.$$

$$m = D(F(\lambda_{kj}, .), \Omega_0(\lambda_{kj}), 0) = d_{kj} + D(F(\lambda_{kj}, .), B_\rho(0), .) \text{ for }$$

$j = 1, 2$ and $\rho > 0$ sufficiently small, where

$$d_{kj} = D(F(\lambda_{kj}, .), \Omega_0(\lambda_{kj}) \setminus \overline{B}_\rho(0), 0).$$

Furthermore $d_{11} = 0 = d_{p2}$ and $d_{k2} = d_{k+1,1}$. Hence

$$\sum_{k=1}^{p-1} d_{k+1,1} + \sum_{k=1}^p D(F(\lambda_{k1}, .), B_\rho(0), 0)$$

$$= pm = \sum_{k=1}^{p-1} d_{k2} + \sum_{k=1}^p D(F(\lambda_{k2}, .), B_\rho(0), 0)$$

and therefore

$$\sum_{k=1}^p [D(F(\lambda_{k2}, .), B_\rho(0), 0) - D(F(\lambda_{k1}, .), B_\rho(0), 0)] = 0.$$

This evidently implies that we have an even number of jumps in the degree. Since we have one at λ_0, and since the jumps occur only at characteristic values of odd algebraic multiplicity, the theorem is proved.

$\square$

Chapter 9

EXERCISES AND HINTS

Every exercise is not answered here. Readers will learn best if they make a serious attempt to find their own answers before peeking at these hints.

1. Let X be a Banach space. $A : X \to X$ is linear and continuous, $b \in X$. For every λ in the spectrum of A holds $|\lambda| < 1$. Let $F : X \to X$ be defined by

$$F(x) = Ax - b.$$

Then there exists a unique $\hat{x} \in X$, $F(\hat{x}) = \hat{x}$. $\hat{x} = \lim x_n$, $x_n = F(x_{n-1}), x_0 \in X$. If A has an eigenvalue λ with $|\lambda| \geq 1$, then (x_n) does not converge for every $x_0 \in X$.

(hint : (a) Proof :

$$
\begin{aligned}
F(x) &= Ax - b \\
F^2(x) &= A^2 x - Ab - b \\
F^3(x) &= A^3 x - A^2 b - b \\
&\;\;\vdots \\
F^n(x) &= A^n x - A^{n-1} b - \ldots - Ab - b
\end{aligned}
$$

129

$$\|F^n(x) - F^n(y)\| \;=\; \|A^n x - A^n y\| \le \underbrace{\|A^n\|}_{K-n}.\|x - y\|.$$

Now it is clear that if $\Sigma K_n < \infty$ by the Weisinger theorem (Theorem - 1.2). It is sufficient to show that

$$\lim \ \sup \sqrt[n]{\|A^n\|} = \lim_n \ \sup \sqrt[n]{K_n} < 1$$

(Spectrum is a compact subset of A.)

(b) If λ is a eigenvalue with $|\lambda| \ge 1$, then $\exists 0 \ne x \in X$ s.t. $Ax = \lambda x$, $\|F^n 0 - F^n x\| = \|A^n 0 - A^n x\| = |\lambda|^n \|x\|$.

So it does not converges in $|\lambda| \ge 1$.)

2 Let (X, d) be a complete metric space, $\psi : [0, \infty) \to [0, \infty)$ continuous, $\psi(t) < t$, if $t > 0$. Let $F : X \to X$, such that for $x, y \in X$

$$d(F(x), F(y)) < \psi(d(x, y)).$$

Then T has a unique fixed point $\hat{x}$, and $\hat{x} = \lim F(x_n)$, $x_n = F(x_{n-1})$, $x_0 \in X$.

(hint : Proof :

$$\begin{aligned} C_n &= d(F^{(n+1)}(x), F^n(x)) \\ &\le \psi(d(F^n(x), F^{n-1}(x))) \\ &= \psi(C_n - 1) \le C_{n-1}. \end{aligned}$$

Since $C_n \ge 0$, C_n is decreasing sequence.

$$\exists \lim C_n = C \ge 0$$

$$\Rightarrow C \le \psi(C) \Rightarrow C = 0.$$

Next, show $(F^n x)$ is a Cauchy sequence.

If not $\Rightarrow \exists \epsilon > 0$ and $\exists (\eta_k), (m_k)$ s.t.

1. $m_{k+1} > n_{k+1} > m_k > n_k$.

Exercises
131

2. $d(F^{m_k+1}x, F^{n_k}x) < \epsilon$.

3. $d(F^{m_k}x, F^{n_k}x) \geq \epsilon$.

4. $d(F^{m_k-1}x, F^{n_k}x) < \epsilon$.

So,

$$
\begin{aligned}
\epsilon \leq d(F^{m_k}x, F^{n_k}x) \quad &\leq d(F^{m_k}x, F^{m_k-1}x) \\
&\quad + d(F^{m_k-1}x, F^{n_k}x) \\
&\leq C_{m_k-1} + \epsilon \\
&\Rightarrow \lim d(F^{m_k}x, F^{n_k}x) = \epsilon. \\
d(F^{m_k}x, F^{n_k}x) \quad &\leq d(F^{m_k}x, F^{m_k+1}x) \\
&\quad + d(F^{m_k+1}x, F^{n_k+1}x) \\
&\quad + d(F^{n_k+1}x, F^{n_k}x) \\
&\leq C_{m_k} + \psi(d(F^{m_k}x, F^{n_k}x)) + C_{n_k}.
\end{aligned}
$$

Taking limit, we get

$$
\epsilon \leq 0 + \psi(\epsilon) + 0 \Rightarrow \epsilon \leq \psi(\epsilon)
$$
$$
\Rightarrow \epsilon = 0 \quad \text{a contradiction}
$$

therefore $(F^n x)$ is Cauchy sequence.

Since X is complete, $\exists \tilde{X} = \lim F^n x$, and
$F(\tilde{X}) = \lim F^{n+1}(\tilde{X}) = \tilde{X} \Rightarrow \tilde{X}$ is a fixed point.

Now it remains to show $\tilde{X}$ is unique. Let $\tilde{X}, \tilde{X}$ be two fixed points. Then

$$
\begin{aligned}
d(\tilde{X}, \tilde{X}) &= d(F(\tilde{X}), F(\tilde{X})) \\
&< \psi(d(\tilde{X}, \tilde{X})) < d(\tilde{X}, \tilde{X})
\end{aligned}
$$

$\Rightarrow d(\tilde{X}, \tilde{X}) = 0 \Rightarrow \tilde{X} = \tilde{X} \Rightarrow$ Uniqueness)

3   Let P be a metric space. For each p let $F_p : X \to X$ be a k-contraction. $0 < k < 1$. Let for fixed $p_0 \in P$ and for all $x \in X$

$$
\lim_{p \to p_0} F_p(x) = F_{p_0}(x).
$$

132   *Nonlinear Functional Analysis*

Then, for each $p \in P, F_p$ has a unique fixed point x_p, and $\lim_{p \to p_0} x_p = x_{p_0}$.

4. Define $f : \mathcal{R} \to \mathcal{R}^2$ by $f(t) = (t, t^2)$ and $g > \mathcal{R}^2 \to \mathcal{R}$ by $g(\xi_1, \xi_2) = \xi_1$ if $\xi_2 = \xi_1^2$ and $g(\xi_1, \xi_2) = 0$ if $\xi_1 \neq \xi_2^2$. Show that the Fréchet derivative $f'(t)$ exists for all $t \in \mathcal{R}$, the *Gâteaux*-derivative $Dg(0,0)$ exists, but $(gof)'(0) \neq Dg(0,0)f'(0)$ (and hence the chain rule fails for *Gâteaux* - derivatives).

(hint : $f : \mathcal{R} \to \mathcal{R}^2$ is defined by $f(t) = (t, t^2)$,

$$g : \mathcal{R}^2 \to \mathcal{R} \ \text{ by } \ g(\xi_1, \xi_2) = \begin{cases} \xi & \text{if } \xi_2 = \xi_1^2 \\ 0 & \text{elsewhere.} \end{cases}$$

To show that $f'(t)$ exists $\forall t \in \mathcal{R}, Dg(0,0)$ exists, but

$$(gof)'(0) \neq Dg(0,0)f'(0).$$

We can easily show $f'(t) = (1, 2t)$. So it exists $\forall t \in \mathcal{R}$. Now

$$\lim_{q \to 0} \frac{g(\tau\xi_1, \tau\xi_2) - g(0,0)}{\tau} = 0 = Dg(0,0)$$

$$(\tau\xi_1)^2 = \tau\xi_2 \Leftrightarrow \tau\xi_1^2 = \xi_2$$

(i) $\xi_1 = 0 \Rightarrow g(\tau\xi_1, \tau\xi_2) = 0$

(ii) $\xi_1 \neq 0$

        (a) $\xi_2 = 0 \Rightarrow g(\tau\xi_1, \tau\xi_2) = 0$

        (b) $\xi_2 \neq 0 \Rightarrow g(\tau\xi_1, \tau\xi_2) = 0$ for $|\tau| < \frac{\xi_2}{\xi_1}$.

So the limit exists i.e. $Dg(0,0)$ exists.

Now $(gof)(t) = g(t, t^2) = t$.

Therefore $(gof)'(0) = 1$. But $Dg(0,0), f'(0) = 0$.)

5. Suppose that X is a real Banach space and $f : X \to \mathcal{R}$ is *Gâteaux* - differentiable on X. If f reaches a relative minimum at $x_0 \in X$ (i.e., there is an $r > 0$ such that $f(x) \geq f(x_0)$ whenever $|x - x_0| \leq r$) show that $f'(x_0) = 0$. If, in addition, f

Exercises

133

is convex, show that f reaches a relative minimum at $x_0 \in X$ if and only if $f'(x_0) = 0$. f is convex

$$\Leftrightarrow \forall x, y \in X \forall \tau \in [0, 1]$$
$$f(\tau x + (1 - \tau)y) \le \tau f(x) + (1 - \tau)f(y).$$

(**hint** : (a) To show $f'(x_0) = 0$

$$0 \le \lim_{t \to 0^+} \frac{f(x_0 + th) - f(x_0)}{t}$$
$$= f'(x_0)h = \lim_{t \to 0^-} \frac{f(x_0 + th) - f(x_0)}{t} \le 0$$

for $th| \le r$ since $f(x) \ge f(x_0)$ for $|x - x_0| \le r \Rightarrow f'(x_0) = 0$.

(b) Suppose $f'(x_0) = 0$

$$\Rightarrow \exists \tau \in (0, 1) \quad : \quad \tau(x_0 + h) + (1 - \tau)(x_0 + th) = x_0$$
$$\Rightarrow h(\tau + t - \tau t) = 0$$
$$\Rightarrow T + t - \tau t = 0$$
$$\Rightarrow \tau = \frac{-t}{1 - t} = \frac{t}{t - 1}$$

$$f(x_0) = f(\tau(x_0 + h) + (1 - \tau)(x_0 + th))$$

i.e., $f(x_0) \le \dfrac{t}{t - 1} f(x_0 + th) - \dfrac{1}{t - 1} f(x_0 + th)$

$$tf(x_0) - f(x_0) \ge f(x_0 + h) - f(x_0 + th)$$

$$\frac{f(x_0 + th) - f(x_0)}{t} \le f(x_0 + h) - f(x_0)$$
$$\Rightarrow f'(x_0) = 0 \le f(x_0 + h) - f(x_0)$$
$$\Rightarrow \min \quad at \quad x_0.)$$

6. : Let $f : [a, b] \times \mathcal{R} \to \mathcal{R}$ be a continuous function. Then by

$$\varphi(x) = \int_a^b \int_0^{x(\tau)} f(\tau, s) ds d\tau$$

a continuously differentiable functional on $C[a,b]$ is defined. In addition $|f(t,s)| \leq c(1+|s|^{p/q})$ on $[a,b] \times \mathcal{R}$, with $c > 0, p > 0$ and $\frac{1}{p} + \frac{1}{q} = 1$, then φ is continuously differentiable on $L_p[a,b]$.

7. Definition

A set $\mathcal{A}$ is called a Banach algebra, if

- $\mathcal{A}$ is a Banach space, and if

- there is defined an associative distributive continuous multiplication of elements of $\mathcal{A}$, i.e.

if $x, y \in \mathcal{A}$, then $x.y \in \mathcal{A}$ and $||x.y|| \leq ||x||.||y||$.

Let $\mathcal{A}$ be a Banach algebra, and $U \subset \mathcal{A}$ an open subset. Let $f, g : U \to \mathcal{A}$ be *Fréchet* differentiable.

(a) Then $h : U \to \mathcal{A}, h(x) = f(x).g(x)$ is Fréchet differentiable. Determine $h'(x)$.

(b) Let $\mathcal{G}\mathcal{A} = \{x \in \mathcal{A}, x^{-1} \text{ exists}, x^{-1} \in \mathcal{A}\}$.

Determine the first and second derivative of $f'(x) = x^{-1}$. Show that $\mathcal{G}\mathcal{A}$ is open in $\mathcal{A}$.

(**hint** :   : A Banach algebra, $U \leq A$ is *Fréchet* differentiable open subset.

(a) $h : U \to A, h(x) = f(x).g(x)$

$$||h(x+k) - h(x) - f'(x)k.g(x) - f(x).g'(x).k||$$
$$= ||f(x+k)g(x+k) - f(x)g(x)$$
$$-f'(x)k.g(x) - f(x).g'(x)k||$$
$$\leq ||f(x+k)g(x+k) - g(x) - g'(x)k||$$
$$+||(f(x+k) - f(x) - f'(x)k)g(x)||$$
$$+||f(x+k) - f(x)g'(x)k||$$
$$\leq ||f(x+k)||||g(x+k) - g(x) - g'(x)k||$$
$$+||f(x+k) - f(x) - f'(x)k||.||g(x)||$$
$$+||f(x+k) - f(x)||||g(x)||.||k||.$$

Exercises

135

$\epsilon > 0$ choose

$\epsilon_1 . \|g'(x)\| \le \epsilon/2$

$\epsilon_2 . \|g(x)\| \le \epsilon/2$

$\epsilon_3 . \|f(x)\| + \epsilon_1 \le \epsilon/3$

$$f \text{ count} \Rightarrow \exists \delta_1 > 0 \text{ s.t. } \|k\| \le \delta$$
$$\Rightarrow \|f(x+k) - f(x)\| \epsilon_1$$
$$F - diff \Rightarrow \exists \delta_2 > 0 \ rms.t. \ \|k\| \le \delta_2$$
$$\Rightarrow \|f(x+k) - f(x) - f'(x)k\| \le \epsilon_1 \|k\|$$
$$gF - diff \Rightarrow \exists \delta_3 > 0 \text{ s.t. } \|k\| \le \delta_3$$
$$\Rightarrow \|g(x+k) - g(x) - g'(x)k\| \le \epsilon_3 \|k\|.$$

Also,

$$\|k\| \le \delta = \min\{\delta_1, \delta_2, \delta_3\}$$
$$\Rightarrow \|h(x+k) - h(x) - f'(x)kg(x) - f(x).g'(x)k\|$$
$$\le (\|f(x)\| + \epsilon_1)\epsilon_3\|k\| + \epsilon_2\|k\|.\|g(x)\|$$
$$+\epsilon_1\|g(x)\|\|k\| \le \epsilon\|k\|$$
$$\Rightarrow h'(x)k = f'(x)k.g(x) + f(x).g'(x)k.$$

(b) $\mathcal{GA} = \{x \in A, x^{-1} \text{ exists}, x^{-1} \in \mathcal{A}\}$. $\mathcal{GA}$ is open.

$$f : \mathcal{GA} \to \mathcal{GA} \text{ by } f(x) = x^{-1}.$$

Let $x \in \mathcal{GA} \Rightarrow x^{-1}$ exists and $x^{-1} \in \mathcal{A} \Rightarrow \|x^{-1}\| \ne 0$.

Consider $\|h\| < \frac{1}{\|x^{-1}\|}$. We shall show $(x - h)^{-1}$ exists.

So, $x - h = x(e - x^{-1}h)$.

Since $\|x^{-1}h\| < 1 \Rightarrow e - x^{-1}h \in \mathcal{GA} \Rightarrow (e - x^{-1}h)^{-1}$ exists

$$\Rightarrow (e - x^{-1}h)^{-1} = \sum_{n=0}^{\infty} (x^{-1}h)^n \in \mathcal{A}.$$

Since $\sum_{n=0}^{\infty} ||x^{-1}h||^n$ is convergent and $\mathcal{A}$ is complete,

$$
\begin{aligned}
\sum_{n=0}^{\infty} (x^{-1}h)^n(\epsilon - x^{-1}h) &= \sum_{n=0}^{6} ((x^{-1}h)^n - (x^{-1}h)^{n+1}) \\
&= \lim_k \sum_{n=0}^{k} ((x^{-1}h)^n - (x^{-1}h)^{n-1}) \\
&= \lim_{k\to\infty} (\epsilon - (x^{-1}h)^{k+1}) = \epsilon
\end{aligned}
$$

$\Rightarrow x - h$ is invertible.

So, $B(x, \frac{1}{||x^{-1}||}) \leq \mathcal{G}\mathcal{A} \Rightarrow \mathcal{G}\mathcal{A}$ is open.

Again, $f(x) = x^{-1}$; take $||h|| < \frac{1}{2||x^{-1}||}$. So,

$$
\begin{aligned}
&||(x+h)^{-1} - x^{-1} + x^{-1}hx^{-1}|| \\
&= ||(x(\epsilon + n^{-1}h))^{-1} - x^{-1} + x^{-1}hx^{-1}|| \\
&= ||(\epsilon + x^{-1}h)^{-1} - (e + x^{-1}h)x^{-1}|| \\
&= ||(\sum_{n=0}^{\infty} (-x^{-1}h)^n - (e + x^{-1}h)x^{-1}|| \\
&= || \sum_{n=2}^{\infty} (-x^{-1}h)^n.x^{-1}|| \\
&= ||(-x^{-1}h)^2 \sum_{n=0}^{\infty} (-x^{-1}h)^{n-2}|| \\
&\leq ||x^{-1}||^3 ||h||^2 \sum_{n=0}^{\infty} (||x^{-1}||\,||h||)^n \\
&= ||x^{-1}||^3 ||h||^2 \frac{1}{1 - ||x^{-1}||\,||h||} \leq 2||x^{-1}||^3 ||h||^2
\end{aligned}
$$

$\epsilon > 0$ given, we can choose $\delta > 0$ s.t.

$$
\delta = \min \left\{ \frac{1}{2||x^{-1}||}, \frac{\epsilon}{2||x^{-1}||^3} \right\}, ||h|| \leq \delta
$$

$$
\Rightarrow ||f(x+h) - f(x) + x^{-1}hx^{-1}|| \leq \epsilon.||h||.
$$

Now

$$
f''(x)(h_1, h_2) = x^{-1}h_1x^{-1}h_2x^{-1} + x^{-1}h_2x^{-1}h_1x^{-1}.)
$$

Exercises 137

8. Consider the eigenvalue problem

$$Ax = \lambda Bx,$$

where A and B are real $(n \times n)$ matrices and where the norm condition $<x, x> = 1$ is satisfied. An important trick is to change this linear problem into a nonlinear one of the form $Tz = 0$, where

$$z = (x, \lambda) \quad \text{and} \quad Tz = (Ax - \lambda Bx, <x, x> -1).$$

Show that the F - derivatives are

$$T'((X, \lambda))(y, \mu) = (Ay - \mu Bx - \lambda By, 2 <x, y>).$$

(**hint :** $Ax = \lambda Bx$, where $A, B \in Mat(n, n; \mathcal{R}), <x, x> = 1, x \in \mathcal{R}^n$. And

$$T(X, \lambda) = (Ax - \lambda Bx, <x, x> -1), (X, \lambda) \in \mathcal{R}^n \times \mathcal{R} = X$$

Let $T : X \to X$. So, F - derivatives are

$$T'((X, \lambda))(y, \mu) = (Ay - \mu Bx - \lambda By, 2 <x, y>)$$

$$T(x, \lambda) = \underbrace{(0, -1)}_{T_o} + \underbrace{(Ax, 0)}_{T_1} + \underbrace{(-\lambda Bx, <x, x>)}_{T_2(x, \lambda)},$$

where $T_0 = $ constant and $T_1 = $ linear operator, where $T_2(tx, t\lambda) = t^2 T_2(X, \lambda) \Rightarrow T_2$ is a homopol on X - into $X \Rightarrow T_2$ is generated by a symmetric bilinear map M, i.e. by $M \in \mathcal{ML}(X, X; X)$ symmetric. So,

$$M((x_1, \lambda_1), (x_2, \lambda_2)) = \frac{1}{2}(-\lambda_1 Bx_2 - \lambda_2 Bx_2, 2 <x_1, x_2>)$$

$$\text{Also,} \quad M((x, \lambda), (x, \lambda)) = T_2(x, \lambda).$$

138   *Nonlinear Functional Analysis*

So we have T_0 (constant) differentiable T_1 (linear operator) is differentiable and T_2 is differentiable. Therefore

$$
\begin{aligned}
T'(x, \lambda)(y, \mu) &= T_0' + T_1' + T_2' \\
&= 0 + (Ay, 0) + 2M((x, x), (y, \mu)) \\
&= (Ay, 0) + (-\lambda By - \mu Bx, 2 < x, y >) \\
&= (Ay - \lambda By - \mu Bx, 2 < x, y >)
\end{aligned}
$$

Also, $T''(x, \lambda)(y_1, \mu_1)(y_2, \mu_2) = 0 + 2M((y_1, \lambda_1), (y_2, \lambda_2))$

since $(Ay, 0)$ is continuous with respect to $x = (-\mu, By_2 - \mu_2 By_1, 2 < y_1, y_2 >)$ this is continuous with respect to x

$$
\Rightarrow T^{(n)}(x, \lambda) = 0 \quad \text{for } n \geq 2.)
$$

9. : Let $F : C[0, \pi] \to C[0, \pi]$ be defined by

$$
F(x)(t) = \frac{2}{\pi} \int_0^\pi k(t, \tau)(x(\tau) + x^3(\tau)) d\tau
$$

where, $k(t, \tau) = a \sin t \sin \tau + b \sin 2t \sin 2\tau, 0 < b < a$. Discuss the set of all solutions of

$$
H(\lambda, x) = \lambda x - F(x) = 0,
$$

where λ is a real parameter.

(**hint** :
$$
\begin{aligned}
\lambda x(t) &= F(x)(t) \\
&= \frac{2}{\pi} a \sin t \int_0^\pi \sin \tau (x(\tau) + x^3(\tau)) d\tau \\
&\quad + \frac{2}{\pi} b \sin 2t \int_0^\pi \sin 2\tau (x(\tau) + x^3(\tau)) d\tau.
\end{aligned}
$$

If x is a solution then x should be linear combination as $x(t) = A \sin t + B \sin 2t$.

$$
\Rightarrow \lambda A \sin t + \lambda B \sin 2t
$$

Exercises

$$= \frac{2}{\pi} a \sin t \int_0^\pi \sin \tau (A \sin \tau + B \sin 2\tau)$$
$$+ (A \sin \tau + B \sin 2\tau)^3) d\tau$$
$$+ \frac{2}{\pi} b \sin 2t \int_0^\tau \underbrace{\sin 2\tau (A \sin \tau + B \sin 2\tau) + (A \sin \tau + B \sin 2\tau)^3}_{T_2}$$
$$= (\frac{2}{\pi} a \sin t)(I_1) + (\frac{2}{\pi} b \sin 2t)(I_2)$$

$$I_1 = \frac{\pi}{2} A + \frac{3\pi}{8} A^3 + \frac{3\pi}{4} AB^2$$

$$I_2 = \frac{\pi}{2} B + \frac{3\pi}{8} B^3 + \frac{3\pi}{4} A^2 B.$$

Since $\sin t$ and $\sin 2t$ are linearly independent in $C[0, \pi]$, we can equate the coefficients of $\sin t$ and $\sin 2t$ to get,

$$\lambda A = aA(1 + \tfrac{3}{4} A^2 + \tfrac{3}{2} B^2) \qquad \text{(i)}$$
$$\lambda B = bB(1 + \tfrac{3}{4} B^2 + \tfrac{3}{2} A^2) \qquad \text{(ii)}$$

Case 1 $\quad A = B = 0 \Rightarrow x(t) = 0$ is a solution for every $\lambda \in \mathcal{R}$.

Case 2 $\quad A = 0, B \neq 0$.

From (ii) $\lambda = b(1 + \frac{3}{4} B^2)$

$$\Rightarrow B = \pm \frac{2}{\sqrt{3}} \sqrt{\frac{\lambda}{b} - 1} \in \mathcal{R} \Leftrightarrow \lambda \geq b$$

$\Rightarrow$ the solution

$$x_{2,3}^{(\lambda)}(t) = \pm \frac{2}{\sqrt{3}} \sqrt{\frac{\lambda}{b} - 1} \sin 2t \text{ for } \lambda \geq b.$$

Case 3 $A \neq 0, B = 0 \Rightarrow$ equation (ii) fulfilled everywhere. Then from (i)

$$\lambda = a(1 + \frac{3}{4} A^2)$$

$$\Rightarrow A = \pm\frac{2}{\sqrt{3}}\sqrt{\frac{\lambda}{a}-1} \in \mathcal{R} \Leftrightarrow \lambda \geq a$$

$$\Rightarrow x_{4,5}^{(\lambda)}(t) = \pm\frac{2}{\sqrt{3}}\sqrt{\frac{\lambda}{b}-1}\sin t.$$

From theorem of local homeomorphism $F : X \to Y$ cont. diff. $F'(x_0)$ is invertible.

$\Rightarrow \exists$ nbd $u(x_0)$ s.t. F is a homeomorphism on $u(x_0)$.

We have

$$\begin{aligned}
H(\lambda, x)(t) &= (\lambda x - F(x))(t) \\
&= \lambda(x(t)) - \frac{2}{\pi}\int_0^\pi k(t,\tau)xtaud\tau \\
&\quad -\frac{2}{\pi}\int_0^\pi x(t,\tau)x^3(\tau)d\tau
\end{aligned}$$

$\Rightarrow H_x'(\lambda, x)$ exists and cont. in x

$$\begin{aligned}
H_x'(\lambda, x)h(t) &= \lambda h(t) - \frac{2}{\pi}\int_0^\tau k(t,\tau)h(\tau)d\tau \\
&\quad -3.\frac{2}{\pi}\int_0^\tau k(t,\tau)x^2(\tau)h(\tau)d\tau \\
H_x'(\lambda, 0)h(t) &= \lambda h(t) - \frac{2}{\pi}\int_0^\pi k(t,\tau)h(\tau)d\tau.
\end{aligned}$$

Since $k(t,\tau) = a\sin t\sin\tau + b\sin 2t\sin 2\tau$,

$$\sin t, \sin 2t \in L_2[0,\pi], \|\sin t\|_2 = \|\sin 2t\|_2 = \sqrt{\frac{\pi}{2}}.$$

Let

$$e_1 = \frac{\sin t}{\sqrt{\frac{\pi}{2}}}, \quad e_2 = \frac{\sin 2t}{\sqrt{\frac{\pi}{2}}}$$

$$\Rightarrow \frac{\pi}{2}\int_0^\pi k(t,\tau)h(\tau)d\tau = a(e_1,h)e_1 + b(e_2,h)e_2$$

$\Rightarrow a$ and b are the eigenvalue of this integral operator

$$\begin{pmatrix} a & 0 \\ 0 & b \end{pmatrix}$$

Exercises

$\Rightarrow H'_x(\lambda, 0)$ is invertible $\Leftrightarrow a \neq \lambda \neq b$.

So, if $a \neq \lambda \neq b$ then from local homeomorphism theorem

$\Rightarrow \exists u_\lambda$ a neighbourhood of 0 in $C[0, \pi]$

$\Rightarrow H(\lambda, 0)$ is a homeomorphism on u_λ since $H(\lambda, 0) = 0 \to 0$

is the unique solution in U_λ.

Case 4

$$A \neq 0, B \neq 0 \Rightarrow \quad \lambda = a(1 + \frac{3}{4}A^2 + \frac{3}{2}B^2)$$

$$\lambda = b(1 + \frac{3}{4}B^2 + \frac{3}{2}A^2)$$

$$\Rightarrow A = \pm\frac{2}{3}\sqrt{\lambda(\frac{2}{b} - \frac{1}{a}) - 1}$$

$$B = \pm\frac{2}{3}\sqrt{\lambda(\frac{2}{a} - \frac{1}{b}) - 1}$$

Since $0 < b < a \Rightarrow (\frac{2}{b} - \frac{1}{a}) > 0$. But we need

$$\lambda(\frac{2}{b} - \frac{1}{a}) \geq 1 \Leftrightarrow \lambda \geq \frac{ab}{2a - b} < a(> 0)$$

and $\lambda(\frac{2}{a} - \frac{1}{b}) \geq 1$ since $\lambda > 0 \Rightarrow \frac{2}{a} - \frac{1}{b} > 0 \Leftrightarrow a < 2b$ we have $\lambda > \frac{ab}{2b-a} > a$.

If $a < 2b$ and $\lambda \geq \frac{ab}{2b-a}$ then

$$x_{6,7,8,9}(\lambda)(t) = \pm\frac{2}{3}\sqrt{\lambda(\frac{2}{b} - \frac{1}{a}) - 1}\sin t$$

$$\pm\frac{2}{3}\sqrt{\lambda(\frac{2}{a} - \frac{1}{b}) - 1}\sin 2t.)$$

10. : Let X be a complete normed space, and $x : [0, 1] \to X$ be continuous. Show that the Riemann integral

$$\int_0^1 x(t)dt$$

is well defined, linear and that

$$\left\| \int_0^1 x(t)dt \right\| \leq \int_0^1 \|x(t)\|dt.$$

Nonlinear Functional Analysis

11. : Let $F : \mathcal{R}^n \to \mathcal{R}^n$ be continuously differentiable and let $F'(x) \neq 0$ on $\mathcal{R}^n$. Then F is a homeomorphism onto $\mathcal{R}^n$ if and only if

$$\lim_{||x|| \to \infty} ||F(x)|| = \infty.$$

(hint : $F : \mathcal{R}^n \to \mathcal{R}^n$ continuously differentiable and det $(F'(x)) \neq 0$ on $\mathcal{R}^n$. Then F is a homeomorphism onto $\mathcal{R}^n$ iff $\lim_{||x|| \to \infty} ||F(x)|| = \infty$.

" $\Rightarrow$" Assume $\exists (x_n)$ in $\mathcal{R}^n$ with $||x_n|| \to \infty$ and $||F(x_n)|| \not\to \infty$.

$\Rightarrow \exists (x_{n_k})$ with $||F(x_{n_k})|| \leq k$

$F(x_{n_k}) \in \overline{B}(0, k)$ since B is compact.

$\Rightarrow F^{-1}(\overline{B}(0, k))$ is also compact.

$\Rightarrow (x_{n_k})$ is bounded.

Which is contradiction to $||x|| \to \infty$.

" $\Leftarrow$ (a) F is continuous.

(b) $F'(x)$ is invertible since $F'(x) \neq 0$.

$\Rightarrow$ We can apply local homeomorphism.

$\Rightarrow \exists$ a nbhd $U(x) : F : U(x) \Rightarrow F(U(x)) \Rightarrow F$ is open.

(c) Suppose $F(x_n) \to y \in \mathcal{R}^n \Rightarrow \exists (x_{n_k})$ is bounded.

Since if the opposite is true then $||x_n|| \to \infty$.

$\Rightarrow ||F(x_n)|| \to \infty$, which contradicts to $F(x_n) \to y$.

$\Rightarrow \exists (x_{n_{k_\ell}}) = F(\lim x_{n_{k_\ell}}) = F(x) \Rightarrow y \in F(\mathcal{R}^n)$.

Since $F(\mathcal{R}^n)$ is open (from (b) and also $F(\mathcal{R}^n)$ is closed $\Rightarrow F(\mathcal{R}^n) = \phi$ or $\mathcal{R}^n$, here it is not empty. So $F(\mathcal{R}^n) = \mathcal{R}^n \Rightarrow F$ is surjective.

(d) F is injective.)

12. : Let (a_{ij}) be a real $(n \times n)$ matrix with $a_{ij} \geq 0$ for all i, j. then A possesses a non-negative eigenvalue. The associated eigenvector can be chosen, such that all coordinates are non-

Exercises

negative. If additionally $\sum_{i=1}^{n} a_{ij} > 0$ for all j, then A possesses a positive eigenvalue.

(PERRON - FROBENIUS).

(hint : $a_{ij} \in Mat(n, n, \mathcal{R})$, $a_{ij} \geq 0$ $\forall$ i, j then A possesses non-negative eigenvalue, $Ax = \lambda x, x \neq 0$,

$x > 0$. If in addition $\sum_{i=1}^{n} a_{ij} > 0 \forall j$ then $\lambda > 0$.

Proof : $K = \{x \in \mathcal{R}^n : x \geq 0, \sum_{i=1}^{n} x_i = 1\}$.

Then k is bounded, closed $\Rightarrow k$ is compact.

Case 1 : $Ax = 0$ for some $x \in k \Rightarrow 0$ is an eigenvalue of A and X is a eigenvector.

Case 2 : $Ax \neq 0$ $\forall$ $x \in k \Rightarrow \sum_{i=1}^{n}(Ax)_i \neq 0$ we define

$$f(x) = \frac{Ax}{\sum_{i=1}^{n}(Ax)_i}, \quad f : k \to k \text{ is continuous })$$

13. : BROUWER's fixed point theorem is equivalent to the following theorem of $POINCAR\acute{E}$ (1886) and BOHL (1904):

Let $f : \mathcal{R}^n \to \mathcal{R}^n$ be a continuous mapping and suppose

$$\exists r > 0 \ \forall \lambda > 0 \ \forall x \in \mathcal{R}^n \ (\|x\| = r \Rightarrow f(x) + \lambda x \neq 0).$$

Then there exists a point $x_0, \|x_0\| \leq r$, such that $f(x_0) = 0$.

14. Construction of Counter Examples to the Brouwer's fixed point theorem.

(a) Construct $U \subset \mathcal{R}$, which is compact, and a continuous $f : U \to U$ without fixed points.

(b) Construct a convex bounded $U \subset \mathcal{R}, f : U \to U$, is continuous, without fixed points.

(c) Find $f : [0, 1] \to [0, 1]$ without fixed points.

(hint :

(a) Compact $U = \{0, 1\}$, $f(0) = 1, f(1) = 0 \Rightarrow f$ is continuous without fixed point (U is also a compact).

144   Nonlinear Functional Analysis

(b) $U = (0,1), f : U \to U$ by $f(X) = \frac{1}{2}X \Rightarrow f$ is continuous without fixed point.

(c) $f : [0,1] \to [0,1]$

$$f(X) = \begin{cases} 0 & \text{for } \frac{1}{2} < X \le 1 \\ 1 & \text{for } 0 \le X \le \frac{1}{2} \end{cases}$$

$\Rightarrow f$ is continuous without fixed point.)

15. : Let X be a separable infinite dimensional Hilbert space with a complete orthogonal system $\{x_0, x_1, x_2, ..., \}$. For

$$x = \sum_{j=0}^{\infty} \alpha_j x_j$$

define

$$F(x) = \sqrt{1 - ||x||^2}.x_0 + \sum_{j=1}^{\infty} \alpha_{j-1} x_j.$$

Show that F maps the closed unit ball into itself continuously with no fixed points. Thus, compactness of F is a necessary hypothesis in the Schauder's fixed point theorem.

(**hint** : X is a separable infinite dimensional Hilbert space, $\{x_0, x_1, ..., ..., \}$ complete orthogonal system.

$$f(x) = \sqrt{1 - ||x||^2}.x_0 + \sum_{j=1}^{\infty} \alpha_{j-1} x_j, ||x_i|| = 1$$

to show that $F : \overline{B}(0,1) \to \overline{B}(0,1)$ is continuous and has no fixed point

$$||F(x)||^2 - 1 - ||x||^2 \sum_{j=1}^{\infty} \alpha_{j-1}^2 = 1,$$

since

$$[X = \Sigma \alpha_j x_j, ||x||^2 = \Sigma \alpha_j^2] \Rightarrow F : \overline{B}(0,1) \to \overline{B}(0,1).$$

Exercises

Next, if $\hat{x} = F(\hat{x}) \Rightarrow ||\hat{x}|| = 1$. So,

$$\hat{x} = \sum_{j=1}^{\infty} \hat{\alpha}_j x_j = \sum_{j=1}^{\infty} \hat{\alpha}_{j-1} x_j$$

$\Rightarrow \hat{\alpha}_0 = 0 \Rightarrow$ by induction $\hat{\alpha}_j = 0 \forall j \in \mathcal{N}$
$\Rightarrow \hat{\alpha}_{j-1} = \hat{\alpha}_j \Rightarrow \hat{x} = 0.$

Finally, F is continuous

$$
\begin{aligned}
||F(x) - F(y)||^2 &= ||(\sqrt{1 - ||x||^2} - \sqrt{1 - ||y||^2})x_0 \\
&\quad + \sum_{j=1}^{\infty} ((x, x_{j-1}) - (y, x_{j-1})||^2 \\
&= (\sqrt{1 - ||x||^2} - \sqrt{1 - ||y||^2})^2 \\
&\quad + \sum_{j=1}^{\infty} (x - y, x_{j-1})^2 \le \epsilon^2 \\
&\forall ||x - y|| \le \delta.
\end{aligned}
$$

Since $\sqrt{1 - ||x||^2}$ is continuous $\Rightarrow F$ is continuous.)

16. : Let X be a partially ordered real Banach space, i.e. there exists a " $\le$", which is compatible with the linear and the topological structure in X, i.e.

1. $x \le y \Rightarrow x + z \le y + z$
2. $\lambda \in \mathcal{R}, \lambda > 0, x \le y \Rightarrow \lambda x \le \lambda y$
3. $x_n \ge 0, \lim x_n = x \Rightarrow x \ge 0$
4. $x_n \ge y_n \ge 0, \lim x_n = 0 \Rightarrow \lim y_n = 0.$

Example : $X = C[0, 1].$

$$x \le y \Leftrightarrow \forall 0 \le t \le 1 x(t) \le y(t).$$

F is called **monotone increasing** iff $x \le y$
$\Rightarrow F(x) \le F(y).$

Let $[x_0, y_0] = \{x = tx_0 + (1-t)y_0, 0 \le t \le 1\}$. **If there exist**
$x_0, y_0 \in X, x_0 \le y_0, x_0 \le F(x_0), y_0 \ge F(y_0)$ **and** $F[x_0, y_0]$ **is**

146   Nonlinear Functional Analysis

relatively compact, then there exists a fixed point $\hat{x}$ of F such that for all $n \in \mathcal{N}$ and for $x_n = F(x_{n-1}), y_n = F(y_{n-1})$

$$x_0 \leq x_1 \leq \ldots \leq x_n \leq \ldots \leq \hat{x} \leq \ldots \leq y_n \leq \ldots \leq y_1 \leq y.$$

(hint : A partially ordered Banach space

$$[x_0, y_0] = \{z \in x : n_0 \leq z \leq y_0\}, F = [x_0, y_0] \to x$$

is continuous and monotonically increasing,

$$x_0 \leq y_0, x_0 \leq F(x_0), y_0 \geq F(y_0), F[x_0, y_0]$$

is relatively compact.

$$X_n := F(x_{n-1}), y_n = F(y_{n-1})$$
$$\Rightarrow \exists x \in [x_0, y_0] : F(\hat{x}) = \hat{x}$$

and

$$x_0 \leq x_1 \leq \ldots \leq \tilde{x} \leq \ldots \leq y_1 \leq y_0.$$

To show $[x_0, y_0]$ is convex : $z_1, z_2 \in [x_0, y_0], t \in [0, 1]$

$$x_0 \leq z_1 \Rightarrow tx_0 \leq tz_1$$
$$x_0 \leq z_2, (1-t)x_0 \leq (1-t)z_2$$
$$\Rightarrow x_0 \leq tz_1 + (1-t)z_2 \leq y_0$$

$[x_0, y_0]$ is closed

$$z \in [x_0, y_0] \Rightarrow F(x_0) \leq F(z) \leq F(y)$$
$$\Rightarrow x_0 \leq F(x_0) \leq F(z) \leq F(y_0) \leq y_0$$
$$\Rightarrow F[x_0, y_0] \leq [x_0, y_0].$$

$F[x, y]$ is relatively compact.

$$CoF[x_0, y_0] \subseteq [x_0, y_0]$$

Exercises

147

$\Rightarrow K = CoF[x_0, y_0] \subseteq [x_0, y_0]$, k is compact

$F(k) \subset F[x_0, y_0] \leq k$

$\Rightarrow F$ is a continuous map on convex compact set. Then we can apply Schauder's fixed point theorem $\Rightarrow \exists \tilde{x} \in k : F(\tilde{x}) = \hat{x}$. Finally, $x_0 \leq x_1 \leq \hat{x} \leq y_1 \leq y_0$ since $\hat{x} \in k \subseteq [x_0, y_0]$. Assume

$$x_{n-1} \leq x_n \leq \hat{x} \leq y_n \leq y_{n-1}$$
$$\Rightarrow F(x_{n-1}) \leq F(x_n) \leq F(\hat{x}) \leq F(y_n) \leq F(y_{n-1})$$
$$\Rightarrow x_n \leq x_{n+1} \leq \hat{x} \leq y_{n+1} \leq y_n$$

$\Rightarrow$ By induction, the assumption is true.)

17. Consider the subsets $B_2 \subset B_3 \subset b_1 \subset C[0, 1]$, defined by

$$\begin{aligned}
B_1 &= \{x : x(0) = 0, x(1) = 1, \ 0 \leq x(t) \leq 1 \text{ in } [0, 1]\} \\
B_2 &= \{x \in B_1 : 0 \leq x(t) \leq \frac{1}{2} \text{ in } [0, \frac{1}{2}] \text{ and} \\
&\quad \frac{1}{2} \leq x(t) \leq 1 \text{ in } [\frac{1}{2}, 1]\} \\
B_3 &= \{x \in B_1 : 0 \leq x(t) \leq \frac{2}{3} \text{ in} [0, \frac{1}{2}] \text{ and} \\
&\quad \frac{1}{3} \leq x(t) \leq 1 \text{ in } [\frac{1}{2}, 1]\}.
\end{aligned}$$

Then $\beta(B_j) = \frac{1}{2}$ for $j = 1, 2, 3$; and $\chi(B_1) = 1, \chi(B_2) = \frac{1}{2}$, and $\chi(B_3) = \frac{2}{3}$.

(**Fact** : $\chi(T) = 2, \beta(U) = 1$ for the unit ball U in an infinite dimensional space.)

(**hint** : Since $B_2 \subset B_3 \subset B_1$ then it is enough to show that $\beta(B_2) = \frac{1}{2}$. We want to show : $\beta(B_2) \geq \frac{1}{2}$.

Assume that $\exists y_1, ..., y_m \in C[0, 1], B_2 \subset \cup_{i=1}^m B(y_i, r)$,

Nonlinear Functional Analysis

$r = \frac{1}{2} - \epsilon < r$. Define

$$x_n(t) = \begin{cases} 0 & \text{if } t \in [0, \frac{1}{2} - \frac{1}{n}] \\ \frac{1}{2} + \frac{n}{2}(t - \frac{1}{2}) & \text{if } t \in [\frac{1}{2} - \frac{1}{n}, \frac{1}{2} + \frac{1}{n}] \\ 1 & \text{if } t \in [\frac{1}{2} + \frac{1}{n}, 1] \end{cases}$$

$$x_n \in B_2.$$

Let i be fixed,

Case I $y_i(\frac{\ell}{2}) \geq \frac{n}{2}$. Since y_i is continuous, then

$$\exists \delta > 0 \ \forall t \in [0, 1] : |t - \frac{1}{2}| \leq \delta$$

$$\Rightarrow |y_i(t) - y_i(\frac{1}{2})| < \epsilon \Rightarrow |y_i(t)| \leq |y_i(\frac{1}{2})| + \epsilon$$

Take n_i such that $\frac{1}{n_i} \leq \delta$

$$
\begin{aligned}
n \geq n_i! \|x_n - y_i\| \ &\geq \ |\underbrace{x_n(\frac{1}{2} - \frac{1}{n})}_{=0} - y_i(\frac{1}{2} - \frac{1}{n})| \\
&= \ |y_i(\frac{1}{2} - \frac{1}{n}) - y_i(\frac{1}{2}) + y_i(\frac{1}{2})| \\
&\geq \ |y_i(\frac{1}{2})| - |y_i(\frac{1}{2} - \frac{1}{n}) - y_i(\frac{1}{2})| \\
&\geq \ \frac{1}{2} - \epsilon = r.
\end{aligned}
$$

$\Rightarrow X_n$ is not in ball $B(y_i, r)$.

On the otherhand

CaseII $y_i(\frac{1}{2}) \leq \frac{1}{2}$

$$
\begin{aligned}
\|x_n - y_i\| \ &\geq \ \|x_n(\frac{1}{2} + \frac{1}{n}) - y_i(\frac{1}{2} + \frac{1}{n})\| \\
&\geq \ 1 - (|y(\frac{1}{2})| + \epsilon) \geq 1 - \frac{1}{2} - \epsilon = \frac{1}{2} - \epsilon = r
\end{aligned}
$$

Exercises

$$\Rightarrow \quad x_n \notin B(y_i, r)$$

$$\Rightarrow \quad n_0 := \max n_i \Rightarrow \text{ for } n \geq n_0 : x_n \notin \cup B(y_i, r$$

$$\Rightarrow \quad \beta(B_j) = \frac{1}{2} \text{ for } j = 1, 2, 3.$$

We know $\beta(B) \leq x(B) \leq \text{diam}(B)$. $\hfill (*)$

$x, y \in B_2$:

(i) $t \in [0, \frac{1}{2}] : 0 \leq x(t) \leq \frac{1}{2}, -\frac{1}{2} \leq -y(t) \leq 0$

$\Rightarrow -\frac{1}{2} \leq x(t) - y(t) \leq \frac{1}{2}$

(ii) If

$$t \quad \left. \begin{array}{c} t \in [\frac{1}{2}, 1] : \quad \frac{1}{2} \leq x(t) \leq 1 \\ -1 \leq -y(t) \leq -\frac{1}{2} \end{array} \right\} -\frac{1}{2} \leq x(t) - y(t) \leq \frac{1}{2}$$

From (i) and (ii)

$$\Rightarrow |x(t) - y(t)| \leq \frac{1}{2} \ \forall t \in [0, 1]$$

$$\Rightarrow \|x - y\| \leq \frac{1}{2} \Rightarrow \quad \text{diam } B_2 \leq \frac{1}{2}$$

$\Rightarrow$ from $(*) \frac{1}{2} \leq x(B_2) \leq \frac{1}{2} \Rightarrow x(B_2) = \frac{1}{2}$.

Also we know $X(B_1) \leq 1$. We assume $\exists M_i$ with $\text{diam}(M_i) \leq 1 - \epsilon < 1 (i = 1, ..., m) B_1 \leq \cup_{i=1}^m M_i$ without loss of generality. Assume $b_1 \cap M_i \neq \phi$. Choose $x_i \in B_1 \cap M_i$

$$\Rightarrow M_i \subseteq \overline{B}(X_i, r) \Rightarrow b_1 \subseteq \cup_{i=1}^m \overline{B}(x_i, r) \supseteq \cup M_i.$$

Since

$$X_i \in B_1 : X_i(0) = 0 \Rightarrow \exists \delta > 0 : \forall t \in [0, 1]$$

$$\forall : t \leq \delta \Rightarrow n_i(t) < \epsilon.$$

Define

$$x(t) = \begin{cases} \frac{t}{\delta} & \text{if } 0 \leq t \leq \delta \\ 1 & \text{if } \delta \leq t \leq 1 \end{cases}$$

$$\Rightarrow x \in B_1 .$$

150   Nonlinear Functional Analysis

But

$$\|x - x_i\| \geq |x(\delta) - x_i(\delta)| > 1 - \epsilon = r$$
$$\Rightarrow x \notin \cup \overline{B}(x_i, r) \Rightarrow x \notin \cup M_i$$
$$\Rightarrow x(B_1) = 1.$$

If we define the function similarly as the case for B, we can show

$$\text{diam}(B_3) \leq \frac{2}{3} \Rightarrow \chi(B_3) \leq \frac{2}{3}.$$

Assume $B_3 \subseteq \cup_{i=1}^m M_i, \text{diam}(M_i) \leq \frac{2}{3} - \epsilon = r < \frac{2}{3}$.
Choose $x_i \in B_3 \cap M_i, \exists \delta \leq \frac{1}{2} \forall i \forall t \in [0,1], t \leq \delta \Rightarrow x_i(t) < \epsilon$.
Define

$$x(t) = \begin{cases} \frac{2}{3} \cdot \frac{t}{\delta} & 0 \leq t \leq \delta \\ \frac{2}{3} & \delta \leq t \leq \frac{1}{2} \\ 1 + \frac{2}{3}t & \frac{t}{2} \leq t \leq 1 \end{cases}$$

$$x \in b_3$$
$$\|x - x_i\| \geq |x(\delta) - x_i(\delta)| > \frac{2}{3} - \epsilon = r$$
$$\Rightarrow x \notin \cup_{i=1}^m M_i.)$$

18. : It is impossible to retract the whole unit ball in $\mathcal{R}^n$ onto its boundary, such that the boundary remains pointwise fixed, i.e. there is no continuous $f : \overline{B}(0,1) \rightarrow \partial \overline{B}(0,1)$, such that $f(x) = x$ for all $x \in \partial \overline{B}(0,1)$. (This result is equivalent to Brouwer's fixed point theorem for the ball.)
(hint : Assume that $\exists f : \overline{B}(0,1) \rightarrow \partial \overline{B}(0,1)$ be continous s.t. $f(x) = x$ for all $x \in \partial \overline{B}(0,1)$ then $g(x) := -f(x), g : \overline{B}(0,1) \rightarrow \partial \overline{B}(0,1)$ is continuous for $x \in \partial \overline{B}(0,1)$ we have $g(x) = -f(x) = -x \neq x$.

Exercises
151

$\Rightarrow g$ does not have a fixed point on the boundary. But it is continuous. So, by Brower's fixed point theorem $\exists$ an open ball i.e. $x \in B(0,1) : g(x) = x$. So, $1 = ||f(x)|| = ||g(x)|| = ||x|| < 1$. Assume $\exists f : \overline{B}(0,1) \to \overline{B}(0,1)$ continuous $f(x) \neq x$ for $x \in \overline{B}(0,1)$, $y = x + a(x)(x - f(x)), a(x) \geq 0$. In our fixed point theorem, it was shown that

$$a(x) = -\frac{(x, x - f(x))}{(x - f(x))^2} + \frac{\sqrt{(x, x - f(x)^2 + (1 - ||x||^2)(x - f(x))^2}}{(x - f(x))^2}$$

$\Rightarrow a : \overline{B}(0,1) \to \mathcal{R}_+$ is continuous. $g(x) := x + a(x)(x - f(x))$ then $g : \overline{B}(0,1)$ and $g(x) = x$ for $x \in \partial B(0,1)$ [if $x \in \partial B \Rightarrow a(x) = 0]$.)

19. Let $F : \mathcal{R}^n \times \mathcal{R}^n$ be continuously differentiable. Let there exist a continuously differentiable function $t \to (\mu(t), x(t))$, such that $\mu : \mathcal{R} \to \mathcal{R}$ is strictly monotone increasing and for all t we have $F(\mu(t), x(t)) = 0$.

Let $d(t) = \det \; F_x'(\mu(t), x(t))$.

(a) If $d(t_1)d(t_2) < 0$ and $t_1 < t_2$, then

$$F(\mu, x) = 0 \tag{*}$$

has a bifurcation point $(\mu(t), x(t))$ with $t_1 < t < t_2$.

(b) If d changes sign at t_0, then $(\mu(t_0), x(t_0))$ is a bifurcation point of $(*)$.

20. *(McDonalds problem) : Given any sandwich, Hamburger, Big Mac, Double Burger, Super Big Mac etc., consisting of bread, ham and cheese. Can this be divided equitably among two people with one slice of a knife, such that each person receives an identical share of bread, ham and cheese?*

In McDonalds dream Dr X and Dr Y appeared and generalized this problem to n dimensions.

If $B_1, ..., B_n$ are bounded, measurable subsets of R^n with

$n \geq 1$. Then there is an $(n-1)$ - dimensional hyperplane which divides all the B_j in half.

References

DEIMLING, K. : " Nonlinear Functional Analysis", Springer, Heidelberg 1985.

DEIMLING, K. : "Nichtlineare Gleichungen und Abbildungsgrade", Springer, Heidelberg 1974.

EISENACK, G.; FENSKE, C.C. : "Fixpunkttheorie", BI, Mannheim 1978.

KRASNOSELSKI, M.A. : "Topological Methods in the Theory of Nonlinear Integral Equations", Pergamon, Oxford 1964.

KRASNOSELSKI, M.A. et al. : "Approximate Solution of Operator Equations", Wolters - Noordhoff, Groningen 1972.

PIMBLEY, G.H. : " Eigenfunction Branches of Nonlinear Operators and their Bifurcations", Springer, Heidelberg 1969.

RIEDRICH, Th. : " Vorlesungen *über* nichtlineare Operatorengleichungen", Teubner, Leipzig 1976.

ZEIDLER, E. : "Nonlinear Functional Analysis and its Applications I : Fixed - Point Theorems", Springer, Heidelberg 1986.

INDEX

A

Algorithm 44
Antipodal theorem 92, 99
Antipodal points 96
Asymptotic test 39

B

Banach space 9, 10, 11, 12, 14, 18, 22, 23, 26, 28, 30, 33, 39, 44, 47, 60, 65, 66, 70, 73, 87, 88, 89, 101, 102, 105, 110, 113, 123
Banach fixed point 2, 7, 8, 53, 74, 114
Bifurcation 105, 109, 110, 112, 113, 115, 123
Bilinear map 22, 30, 31
Bijective 30
Bilinear operator 31
Bounded 24, 70, 72
Brouwer's fixed point 55, 57, 59, 63, 64, 82
Brouwer's degree 77, 78, 80, 84, 87, 89, 90
Borsuk 92, 96
Borsji-Ulam 95

C

Contraction 1, 6, 9, 48
Complete metric-space 2, 4, 7, 8
Cauchy sequence 3,4, 43
Construction 3
Compact metric space 5
Convergent 5, 39, 67
Closed ball 8, 12, 14
Continuous 19, 21, 26, 29, 34, 36
Chain rule 23, 26, 32, 37
Complex space 55
Convex 60, 63, 74, 80
Convex hull 60
Compact operator 99

D

Derivative 17
Dual space 17
Diameter 65
Darbo 74
Deimling 78

E

Error estimate 3
Enumeration 56

Euclidean space 57
Eigen vector 81, 111, 117
Eigenvalue 81, 109, 111, 113, 119, 120

F

Fixed point 1, 2, 6, 14, 55, 64, 75, 80
Fixed point index 119
Fréchet derivative 17, 18, 19, 22, 23
Fréchet differentiable 20, 21, 23, 25, 26

G

Gauß theorem 59
Gâteaux derivative 17, 18, 19
General Topology 61

H

Halm-Banach theorem 57
Hedgehog theorem 83
Hoelder condition 53
Homeomorphism 12, 25, 49, 80, 81, 88, 102, 111, 114
Homotopy 83, 89, 98, 103, 116, 124

I

Inequality 12, 43
Injective 2, 12
Initial value problem 8
Invertible 49, 50
Implicit problem 51
Isomorphic 22
Isomorphism 30, 51

K

k-contraction 1, 2, 10, 74
Kuratowski 65

L

Lebesgue measure 84
Leray-Schauder degree 87, 89, 90, 99, 101, 110
Linear map 18, 20, 30
Lipschitz continuous 1, 8, 9, 10, 28, 44, 73
Lipschitz condition 7
Lipschitz constant 9, 10, 73
Local Homeomorphism 11, 26

M

Maping 1
Mean value theorem 26, 27, 31, 48, 52

Index

157

Measure of non-compactness
65, 66, 96
Metric spaces 1
Multilinear mapping
28

N

Neighourhood 5, 12,
47, 49, 51, 53, 113,
115, 119, 117, 124,
125, 126, 127
Newton iterates 42
Newton's method 39,
41, 44, 53
Nonlinear 7
**Non-linear functional
analysis** 78
Norm 30
Normed space 1

O

**Ordinary Differential
Equation** 81
Operator 8, 21

P

Partial derivatives 36
Partial differentiable
36
Picard - Lindelof 8
Proper 73

R

Real-valued function
34, 41
Resolvant operator
9, 10, 11

S

Sadovski 74
Sard's lemma 84,
94
Schauder's fixed point
60, 63, 74, 75
Sequence 9, 39
Strict 1
Surjective maps 83

T

Taylor formula 34
**Topological compo-
sition** 100

U

Unique fixed point
48
Uniqueness 3

V

**Volterra integral equa-
tion** 7, 8

W

Weakly continuous
19
**Weierstrass approx-
imation theorem**
57